C0-DXE-788

	SEALIONS	SEABIRD COLONY	BISON		GRIZZLY BEARS	WORLD OF REPTILES
	MONKEY HOUSE	WORLD				

ANIMALS

Zoo Center
Elephants
Tapirs

Monkey House
Tamarins
Marmosets

The Bronx Zoo
OFFICIAL GUIDE

WCS
WILDLIFE CONSERVATION SOCIETY

Copyright © 1999 Wildlife Conservation Society

2300 Southern Boulevard
Bronx, NY 10460

All rights reserved.
Reproduction of this book, in whole or part,
without written permission of the publisher is prohibited.

For more information about the Wildlife Conservation Society,
visit www.wcs.org

The Bronx Zoo Official Guide

ISBN 0-9632064-3-5

1. Bronx Zoo–Guidebooks. I. Squire, Ann. II. Wildlife Conservation Society.
III. Title.

Writer:	Ann Squire
Photo editor / Coordinator:	Diane Shapiro
Designer:	Lisa Feldman
Production:	MM Design 2000, Inc.
Printer:	Paul-Art Press

Printed and bound in the U.S.A.

COVER: MANDRILL

PAGE 1: LOWLAND GORILLA

PAGES 2-3: LION HOUSE SCULPTURE

PAGE 8: RAINEY GATE

TABLE OF CONTENTS

- **9** INTRODUCTION
- **10** MAP OF THE BRONX ZOO

PART ONE: BRONX ZOO EXHIBITS*

- **15** ZOO CENTER
- **16** ASTOR COURT
- **17** MONKEY HOUSE
- **19** AQUATIC BIRD HOUSE
- **21** RUSSELL B. AITKEN SEA BIRD COLONY
- **22** BIG BIRDS
- **22** PELICANS AND GIBBONS AT COPE LAKE
- **24** FLAMINGO POND
- **24** BIRDS OF PREY
- **26** WILD HORSES AND GUANACOS
- **27** CHILDREN'S ZOO
- **30** WORLD OF REPTILES
- **31** MOUSE HOUSE
- **32** JOHN PIERREPONT WILDFOWL MARSH
- **33** CONGO GORILLA FOREST
- **42** BABOON RESERVE AT SOMBA VILLAGE
- **48** CARTER GIRAFFE BUILDING
- **50** AFRICA
- **52** JUNGLEWORLD
- **60** WILD ASIA
- **67** WORLD OF DARKNESS
- **70** BIG BEARS
- **71** HIMALAYAN HIGHLANDS
- **72** NORTHERN PONDS
- **74** MEXICAN WOLVES
- **74** PERE DAVID DEER
- **76** WORLD OF BIRDS
- **78** BISON RANGE

PART TWO: BEHIND THE SCENES

- **82** KEEPING THE ANIMALS HEALTHY
- **85** MAKING IT ALL WORK
- **88** EDUCATION
- **91** SCIENCE RESOURCE CENTER
- **93** INTERNATIONAL CONSERVATION
- **94** BRONX ZOO BREEDING PROGRAMS
- **97** HISTORY OF THE ZOO AND WCS
- **103** WCS PARKS THROUGHOUT NEW YORK CITY
- **107** MEMBERSHIP, MAGAZINE, WEB SITE
- **109** GENERAL INFORMATION
- **110** INDEX OF ANIMALS

*This guide, like the zoo itself, is organized around the Zoo Center/Astor Court hub. To locate an animal or habitat on the zoo map, *refer to the letter and number key after section titles.*

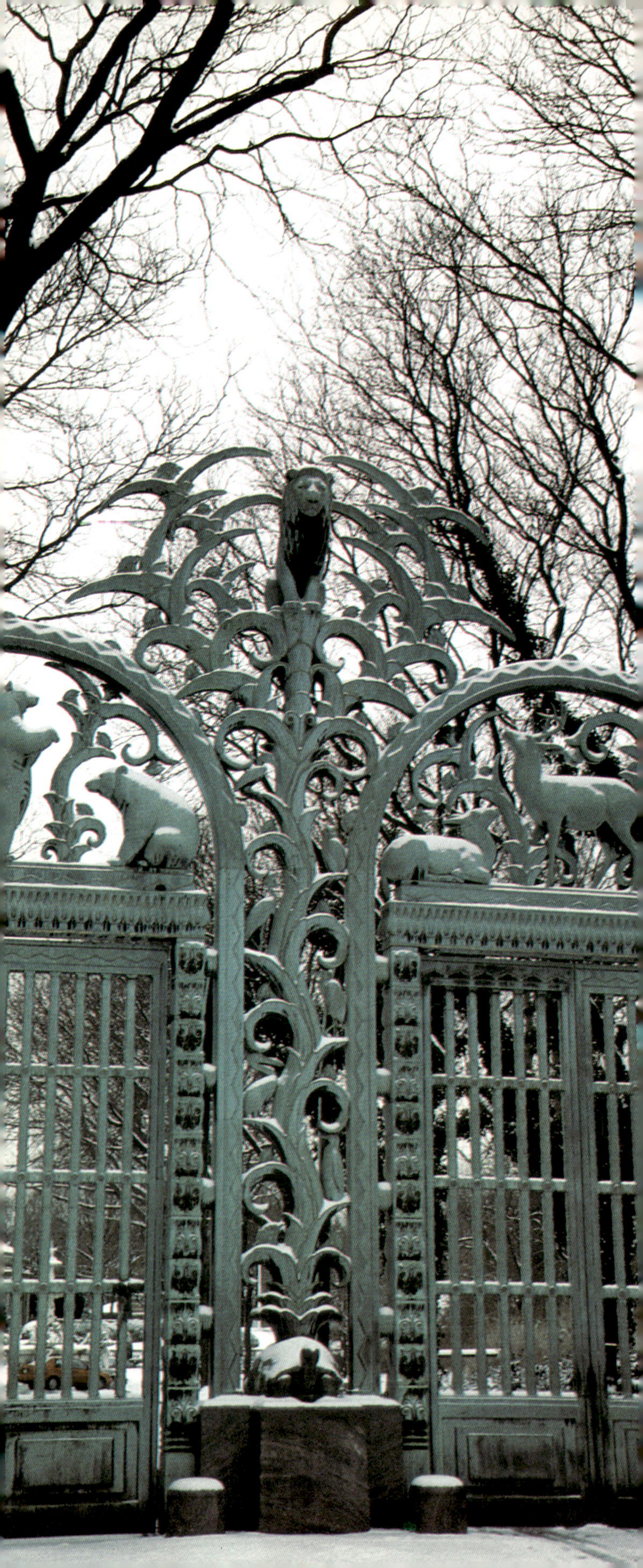

INTRODUCTION

The biggest animal at the Wildlife Conservation Society's renowned Bronx Zoo is a 9,750-pound elephant. The smallest is probably a baby walking-stick. After a good meal, it may weigh 1/1000th of an ounce.

Most of our 7,000 animals are native New Yorkers, hatched or born at the zoo, although their homelands include every continent. For the inquisitive zoo goer, a visit to see the animals is like taking a trip around the world.

Shorter journeys can also be made at the three other zoos and aquarium operated by the nonprofit Wildlife Conservation Society (WCS) for New York (please see page 103 of the guide). Each WCS wildlife center has specialties found nowhere else and each offers particular insights into the fast-vanishing world of nature.

Every year, throughout the world, about 35 million more acres of tropical rain forest are being cut for timber or converted to agriculture; more than 19 million trees are felled every day. Marshes are being drained, savannas plowed, and wild animals are being killed for their meat or just for their skins or teeth.

Although Americans are among the largest users of the world's natural resources, they are also the world's most powerful and serious conservationists. Just a few, however, realize that the Bronx Zoo is home to one of the world's largest wildlife conservation programs. The Wildlife Conservation Society operates more than 300 conservation projects in 52 nations. Its great animal collections at its New York facilities inspire and urge on WCS staff and supporters. Revenue generated at the facilities helps support conservation action in the field.

This unified approach effectively makes a visit to the Bronx Zoo and its sister institutions an act of conservation, and each wonderful animal a special ambassador for its species and for the preservation of nature.

WILLIAM CONWAY
Senior Conservationist

Bronx Zoo Map

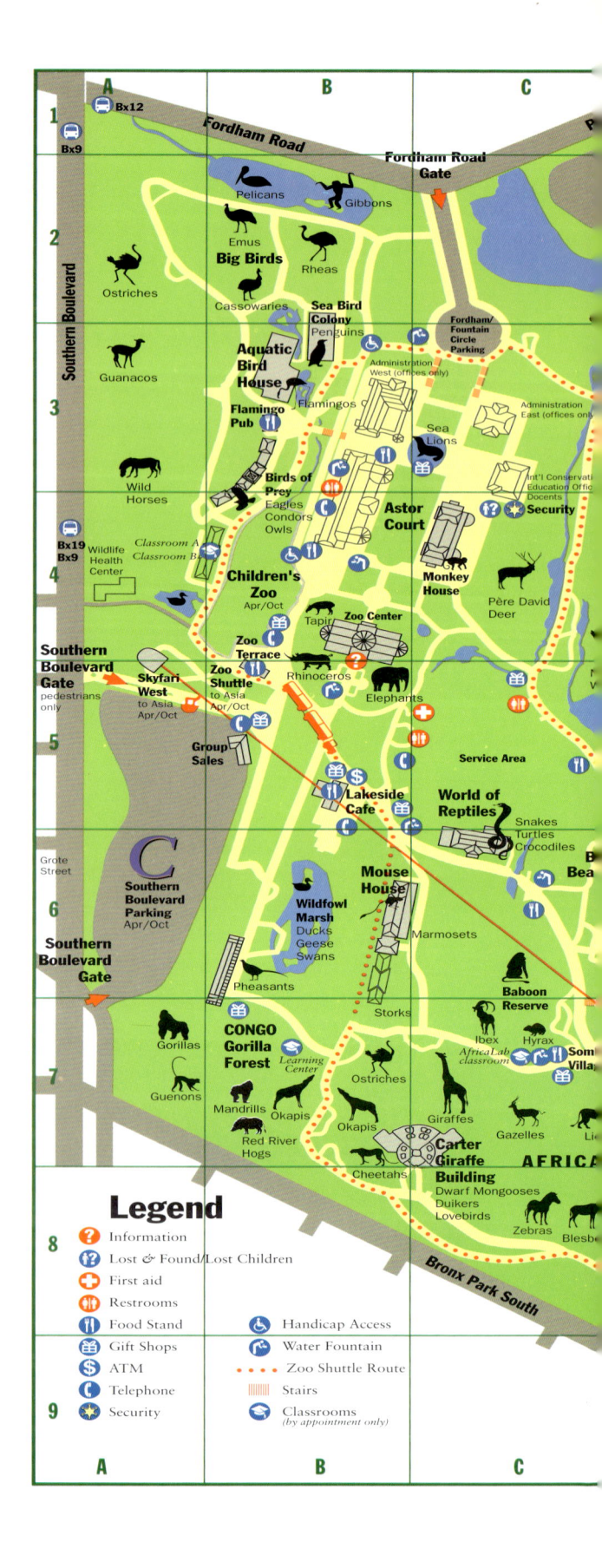

Grid A

1: Bx12, Bx9

2: Southern Boulevard, Ostriches

3: Guanacos

4: Wild Horses, Bx19 Bx9, Wildlife Health Center

5: Southern Boulevard Gate (pedestrians only), Skyfari West to Asia Apr/Oct

6: Grote Street, Southern Boulevard Parking Apr/Oct, Southern Boulevard Gate

7: Gorillas, Guenons

Grid B

1: Fordham Road

2: Pelicans, Gibbons, Emus, Big Birds, Rheas, Cassowaries, Sea Bird Colony

3: Aquatic Bird House, Penguins, Flamingo Pub, Flamingos, Administration West (offices only)

4: Birds of Prey (Eagles, Condors, Owls), Classroom A, Classroom B, Children's Zoo Apr/Oct, Astor Court

5: Zoo Terrace, Zoo Center, Zoo Shuttle to Asia Apr/Oct, Tapir, Rhinoceros, Elephants, Group Sales

6: Lakeside Cafe, Mouse House, Wildfowl Marsh (Ducks, Geese, Swans), Marmosets, Pheasants, Storks

7: CONGO Gorilla Forest, Learning Center, Mandrills, Okapis, Red River Hogs, Ostriches, Okapis, Cheetahs, Carter Giraffe Building (Dwarf Mongooses, Duikers, Lovebirds)

8: Bronx Park South

Grid C

1: Fordham Road Gate

2: (lake)

3: Fordham/Fountain Circle Parking, Administration East (offices only), Sea Lions

4: Int'l Conservation Education Office, Docents, Security, Monkey House, Père David Deer

5: Service Area

6: World of Reptiles (Snakes, Turtles, Crocodiles), Bea...

7: Baboon Reserve, Ibex, Hyrax, AfricaLab classroom, Som... Villa, Giraffes, Gazelles, Li..., AFRICA, Zebras, Blesbo...

Legend

- ❓ Information
- Lost & Found / Lost Children
- ➕ First aid
- Restrooms
- 🍴 Food Stand
- Gift Shops
- $ ATM
- C Telephone
- ★ Security
- ♿ Handicap Access
- Water Fountain
- ••• Zoo Shuttle Route
- ||||| Stairs
- Classrooms (by appointment only)

Bronx Zoo Map

Parkway (D)

IRT #2 (F1)

BxM11 — Entrance (E2)

B — Bronx Parkway Parking (E2-3)

Exit (F3)

Bronx Parkway Gate (D3)

Cranes (D3-4)

World of Birds (D4)
- Hornbills
- Parrots
- Birds of Paradise

Bronx River (E4-5)

Bronx River Parkway (F4-5)

Northern Ponds (D5)
- Swans
- Diving Ducks
- Cranes

JungleWorld Road Parking (D5-6)

Polar Bears (D6)

Himalayan Highlands (D6)
- Snow Leopards
- Red Pandas
- Cranes

World of Darkness (D7)
- Bats
- Leopard Cats
- Mole Rats
- Owl Monkeys

A — Parking Asia Apr./Oct. (D7)

Skyfari East to Zoo Center Apr/Oct (D7)

Zoo Shuttle to Zoo Center Apr/Oct (D8)

Nyalas (D7)

Oryxes (D8)

Camel Rides (D8)

classroom C (D8)

Bengali Express Monorail May/Oct (E8)

Asia Plaza (D8)

JungleWorld (D9)
- Gibbons
- Monkeys
- Leopards
- Crocodiles
- Insects

Jungle Lab classroom (E8)

Asia Gate pedestrians only (D9)

Q44 / **IRT #2 & #5** (D9)

WILD ASIA — May/Oct. (E-F, 5-9)

- Sambar Deer
- Indian Rhinoceros
- Tufted Deer
- Asian Elephants
- Tahrs
- Babirusas
- Red Pandas
- Siberian Tigers
- Axis Deer
- Eld's Deer
- Blackbuck
- Gaur
- Barasingha
- Formosan Sika Deer

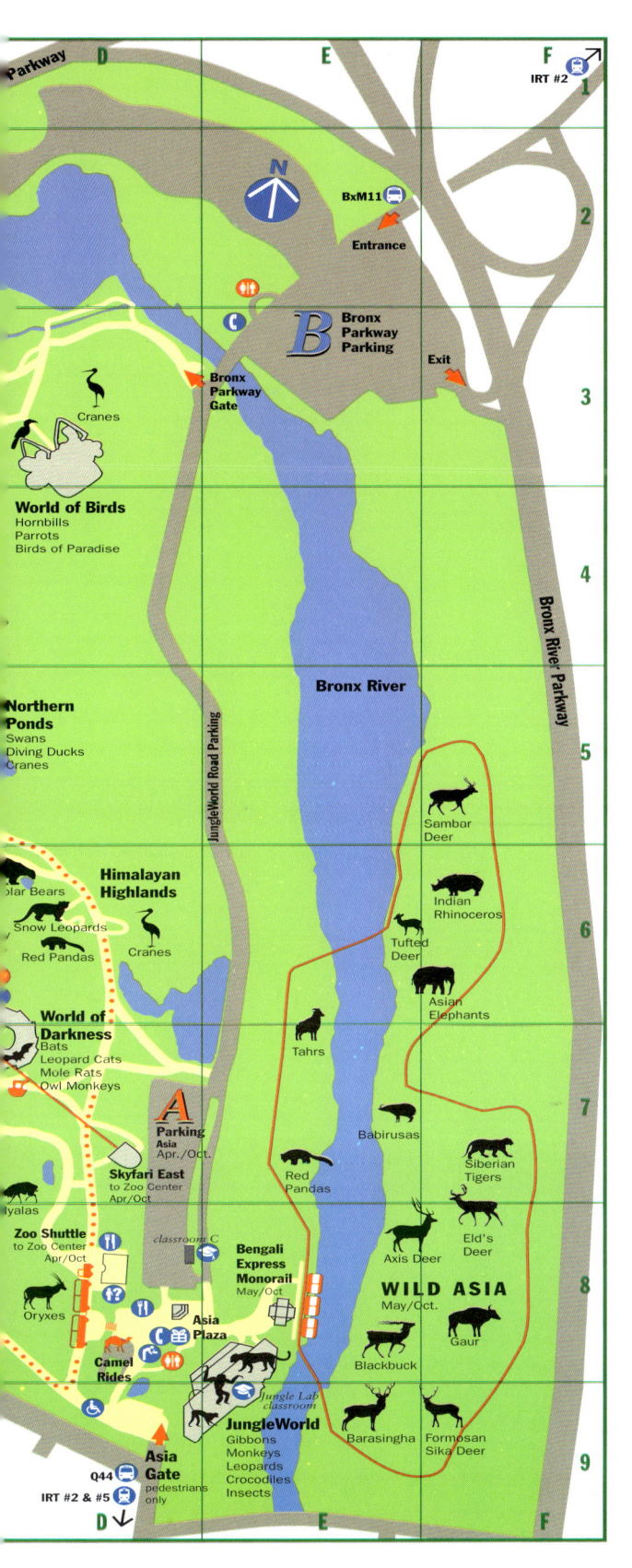

TO LOCATE AN ANIMAL OR EXHIBIT ON ZOO MAP, REFER TO THE LETTER AND NUMBER KEY AFTER SECTION TITLES. (SEE PAGE 15 FOR EXAMPLE)

Part One

BRONX ZOO EXHIBITS

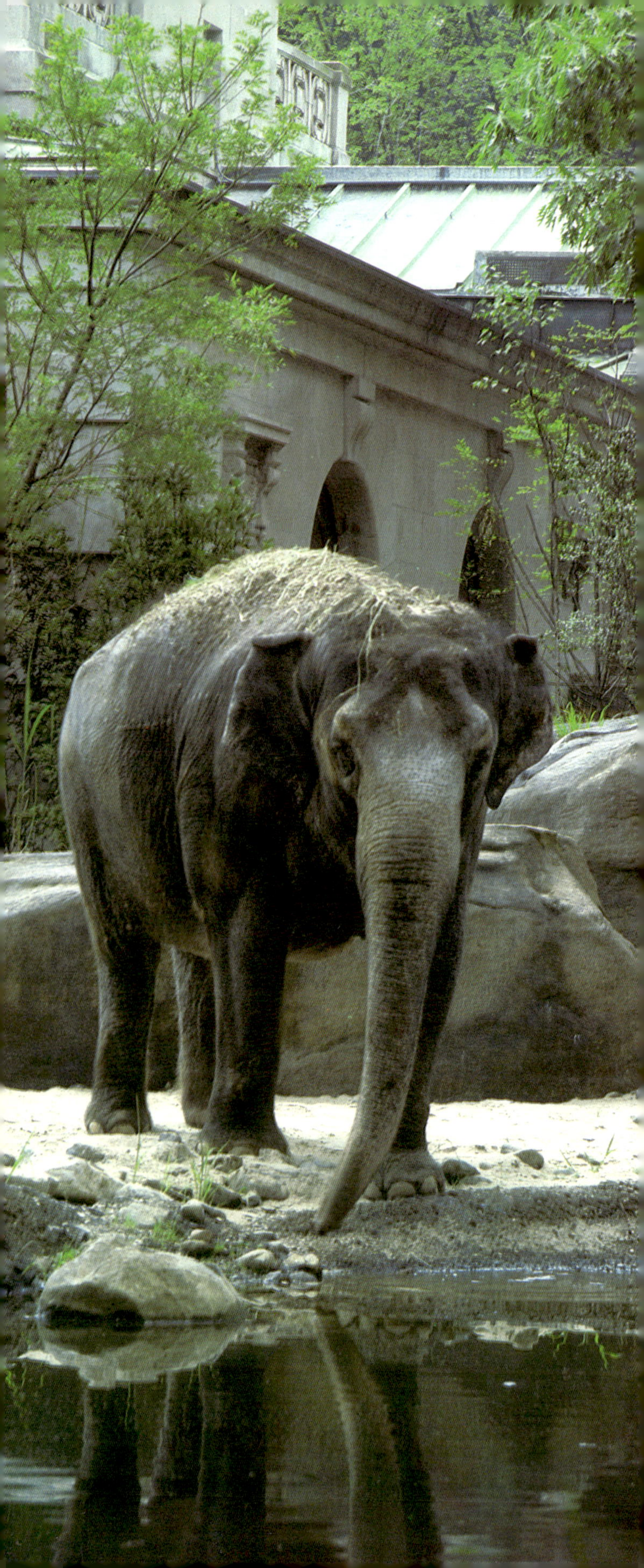

LETTER AND NUMBER KEY ↓

ZOO CENTER (B4)

At the south end of Astor Court stands the Keith W. Johnson Zoo Center. This Beaux Arts building, first opened in 1908, provides indoor and outdoor habitats for Asian elephants, African black rhinos, and Malayan tapirs.

Asian elephants can be distinguished from their African counterparts by their smaller ears and smaller overall size and weight. In the Asian elephant, usually only the males have tusks, while in African elephants, both males and females may have them. The elephant's most amazing feature is certainly its trunk—actually an elongated and highly specialized nose. The elephant uses its trunk to breathe, smell, suck up water (which is then squirted into the mouth), knock down trees, and pick up tiny objects (by employing a finger-like projection at the tip of the trunk).

Asian elephants are much more endangered than the African species. Fewer than 50,000 remain, and their habitats are becoming increasingly fragmented as a result of human population growth and development.

> The monumental Zoo Center is the winter home of Asian elephants and tapirs of Southeast Asia and the year-round home of the zoo's information desk.

African black rhinos, like all other rhinoceros species, are seriously endangered. One reason is widespread demand for rhino horn, which is thought by some to function as a medicine. The horn is actually composed of greatly modified hairs, which are fused together into a strong, solid mass. Many traditional Asian medicines make use of powdered rhino horn, although there is no evidence that it has any effect. Because rhino horn can bring so much on the black market, the incentive for illegal poaching of rhinos remains high.

↑ AFRICAN BLACK RHINO
↑ MALAYAN TAPIR
← ASIAN ELEPHANT

Malayan tapirs are solitary animals that spend their nights feeding on leaves and aquatic plants and their days asleep, hidden

TO LOCATE AN ANIMAL OR EXHIBIT ON ZOO MAP, USE LETTER AND NUMBER KEY AFTER SECTION TITLES. (SEE EXAMPLE ABOVE)

among the foliage along a river's edge. These inhabitants of Asian tropical forests often spend hours partially submerged in shallow water. The Malayan tapir's most striking feature is its color: the legs, front, and rear of its body are black, with a white saddle in between. The tapir's nose and upper lip are extended into a short, movable trunk. Despite its piglike appearance, the tapir's closest relatives are rhinos and horses.

In the areas surrounding Zoo Center, you'll find places to relax among the beautifully landscaped lawns and gardens. In all, thousands of plants, including many unusual species, have been used in the Zoo Center gardens. At the south side of Zoo Center is the "Rhino Garden," guarded by a pair of larger-than-life bronze Indian rhinos.

Inside Zoo Center you'll find the animals' winter quarters, a visitor information center, and participatory exhibits, videos, and artifacts on rhino and elephant conservation.

ASTOR COURT (B4)

At the center of the zoo's magnificent Astor Court is the Sea Lion Pool. Home to our California sea lions, this habitat replicates the rocky beaches, islands, and tidal pools of the California coastline. The pool itself holds 200,000 gallons of water.

↑ SEA LION
↓ SEA LIONS

California sea lions are supremely adapted

to life in the water. Instead of front feet, the sea lion has large oar-like flippers. The smaller hind flippers are used mainly for steering in the water but can be turned forward to allow the sea lion to walk on all fours on land. Other adaptations to life in the water include a smooth, streamlined body shape and a thick layer of insulating fat.

In the wild, sea lions eat fish and squid and are able to dive up to 600 feet in search of food. Male sea lions can reach weights of over 800 pounds, more than three times the size of the females. With their impressive size and loud roar, it's no wonder these marine mammals are called lions—next to humans, they are said to be the noisiest of all mammals!

As you leave the Sea Lion Pool, take a look at the fantastic animal art surrounding you: Rainey Gate, which stands at the Fordham Road entrance to the zoo; the 17th-century limestone fountain at the center of the parking area; and the animal sculptures gracing the buildings around Astor Court.

MONKEY HOUSE (C4)

Located at the southeast corner of Astor Court, the Monkey House is home to smaller primate species. Marmosets, which live in the tropical forests of South and Central America, are among the world's smallest primates. The pygmy marmoset can fit easily in a person's hand and weighs about three ounces. Marmosets are social animals and live in extended family groups. Although they have excellent vision, they communicate with each other primarily by smell and vocalizations. In their natural habitat, marmosets spend most of their time in the treetops feeding on sap, insects, and fruits.

Tamarins eat a wider range of foods than do marmosets. They also occupy much larger territories, which may be related to their need for a more varied diet.

Marmosets and tamarins generally give birth to twins, although triplets are not uncommon, especially in captivity. Unlike the males of most other primate species, marmoset and

↑ SILVERY MARMOSET

↑ GOLDEN LION TAMARIN

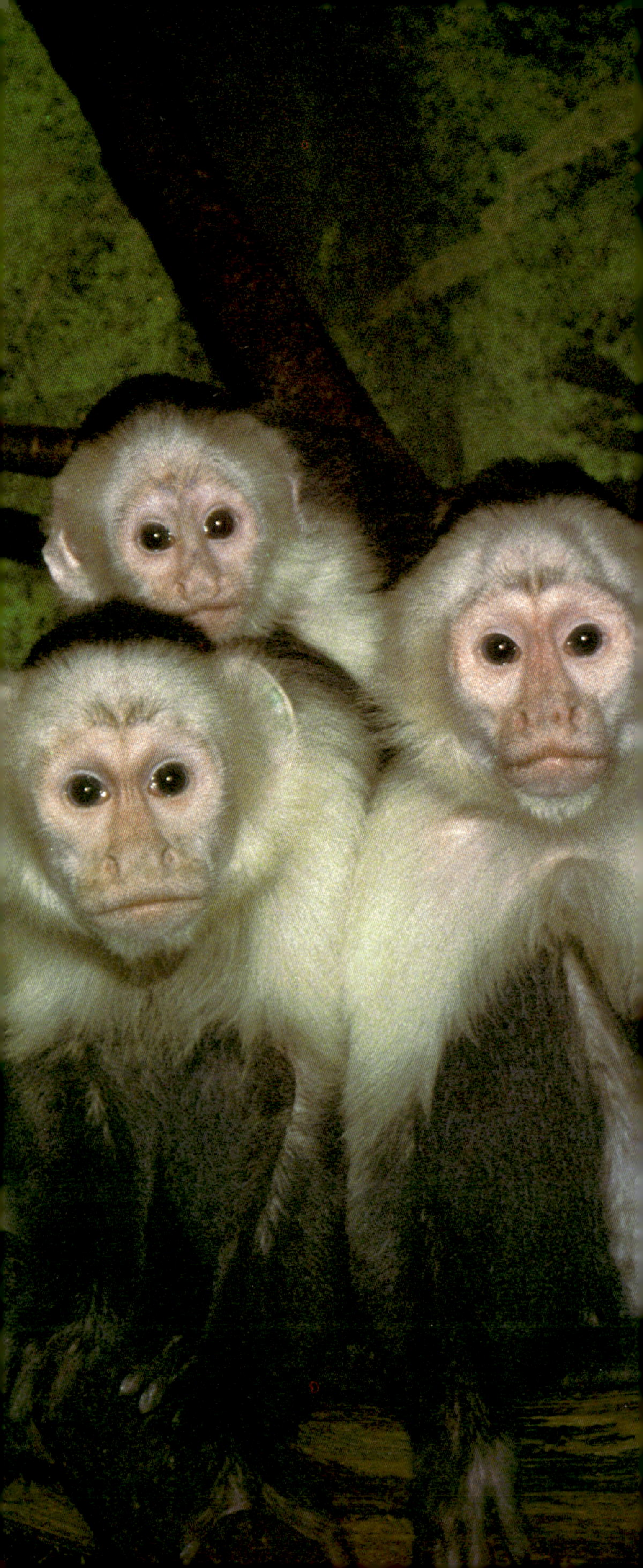

tamarin fathers are very attentive, helping to carry and care for the young.

The Monkey House is also home to capuchin monkeys, which are found in some forested regions of South and Central America. Because of their playfulness, intelligence, and agility, capuchins have long been popular as trained animals. Unfortunately, these traits have also made them targets of the pet trade, which has reduced their numbers in the wild.

AQUATIC BIRD HOUSE (B3)

The Aquatic Bird House contains not just an impressive variety of water birds but an impressive array of habitats, from riverbank to lagoon and from seashore to swamp. With its focus on naturalistic environments, the Aquatic Bird House was one of the first Bronx Zoo exhibits to present animals in the kinds of places they actually live.

One of the highlights is Sea Cliffs, an exhibit that simulates a rugged cliff at ocean's edge. This exhibit is home to a breeding colony of tufted puffins. A glass-fronted aquarium allows you to see these chunky birds nesting in the rocky crevices, bobbing on the water's surface, and using their powerful wings to "fly" underwater.

Some of the most spectacular inhabitants of the Aquatic Bird House are scarlet ibises. In the wild, these birds get their brilliant red color by eating small creatures rich in carotenoids. At the Bronx Zoo these birds, as well as the flamingos just outside, are fed a specially supplemented diet to preserve their color.

↑ SCARLET IBIS
← WHITE-FACED CAPUCHINS

Many of the exhibits, including the Marsh and the Swamp, have no barriers. Lit from above by skylights, they offer excellent opportunities for photographs.

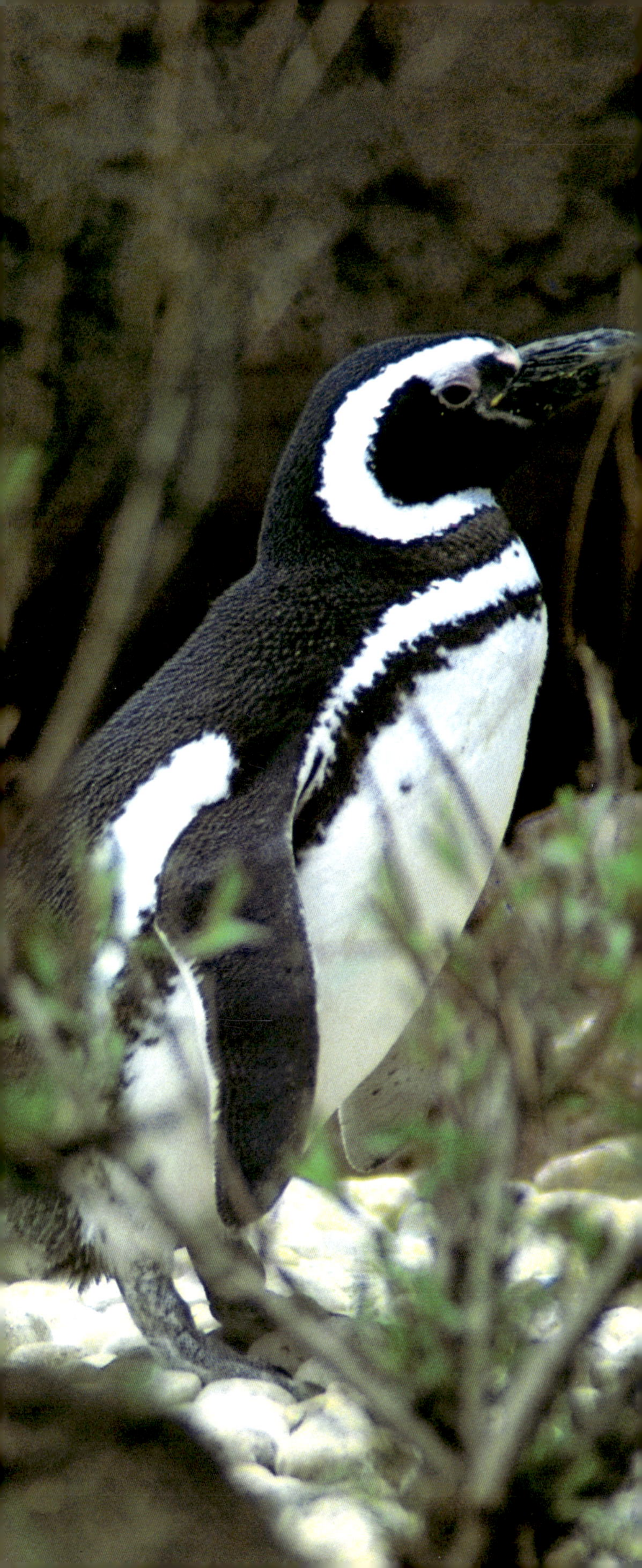

RUSSELL B. AITKEN SEA BIRD COLONY (B2)

The Russell B. Aitken Sea Bird Colony was built to replace the De Jur Aviary after it was destroyed in a fierce storm during the winter of 1995. The Sea Bird Colony now provides a large naturalistic environment for a variety of oceangoing birds.

One of the most striking birds in the aviary is the Inca tern, which you'll recognize by its bright red feet and the long mustache-like white plumes curling backward from the red bill. This gray bird, about the size of a small gull, is highly social, often gathering by the thousands along the rocky shores and sandy beaches of western South America. Like other seabirds, the Inca tern feeds on fish and can sometimes be spotted in the company of cormorants and sea lions as these animals search for food.

Another social bird is the guanay cormorant, a large black-and-white bird with a long, slender neck. Like Inca terns, guanay cormorants inhabit the western coastlines of South America. On the lookout for fish, they fly at considerable heights, often with other birds such as boobies. Once a school of fish has been spotted, the cormorants plummet into the sea to capture their prey. During the breeding season, guanay cormorants gather on islands and headlands in huge nesting colonies that may consist of several hundred thousand birds. Although not currently endangered, guanay cormorants are declining as a result of the commercial fishing industry, which depletes these seabirds' food source, and guano (fertilizer) mining, which disrupts their nest sites.

↑ GUANAY CORMORANT
↑ INCA TERN
← MAGELLANIC PENGUIN

While cormorants and terns spot fish from the air, Magellanic penguins take a more direct approach. These stocky black-and-white birds use their wings, which have been modified into flippers, to propel themselves hundreds of feet underwater in pursuit of their prey. Penguins are the most aquatic of all birds. Among their adaptations to a life at sea are their densely packed water-repellent feathers, a streamlined, torpedo-like shape, and a thick layer of fat to insulate them against the cold ocean waters. Penguins do come out of the water during breeding season, when they nest on land in large colonies. For many years, the Wildlife Conservation Society has sponsored projects designed to conserve the seabirds and other wildlife of coastal South America.

BIG BIRDS (B2)

In the Big Birds area, you'll find just that: very large birds, which are specialized for running rather than flying. The ostrich, probably the best known of the flightless birds, was once common in Africa. Hunting has decimated many ostrich populations and led to the extinction of one ostrich subspecies. However, ostriches are still found in the wild in many parts of east and south Africa. Male ostriches are larger than the females, reaching heights of up to eight feet and weights of almost 350 pounds. The ostrich's foot has only two toes and is highly adapted for running. Ostriches can easily sprint across the African plains at speeds of up to 35 miles per hour.

↑ COMMON RHEA
→ MASAI OSTRICH

The rhea is another large, heavy bird that runs rather than flies. Rheas live in South America, are very social, and live in large flocks. During the breeding season, groups of females move from male to male, laying their eggs in one male's nest after another.

Occupying the flightless bird niche in Australia and New Guinea are emus and cassowaries. These quiet, fruit-eating birds can be fearsome fighters if threatened, and the larger cassowaries of New Guinea are capable of killing a person with their huge, sharp toenails.

PELICANS AND GIBBONS AT COPE LAKE (B2)

From spring through fall, white-cheeked gibbons roam the islands and swing through the trees, while pelicans swim in the waters of this lagoon, which runs along the zoo's northern edge.

Gibbons spend most of their time in the treetops, using their long, strong arms to move rapidly from branch to branch. Of all the primates, they seem the best adapted to arboreal life. In the wild, gibbons live in family groups and, in the early mornings, announce ownership of their territory with loud, strangely beautiful calls. They defend their homes vigorously, chasing away intruders who dare to venture past their territorial boundaries. Gibbons are vegetarians

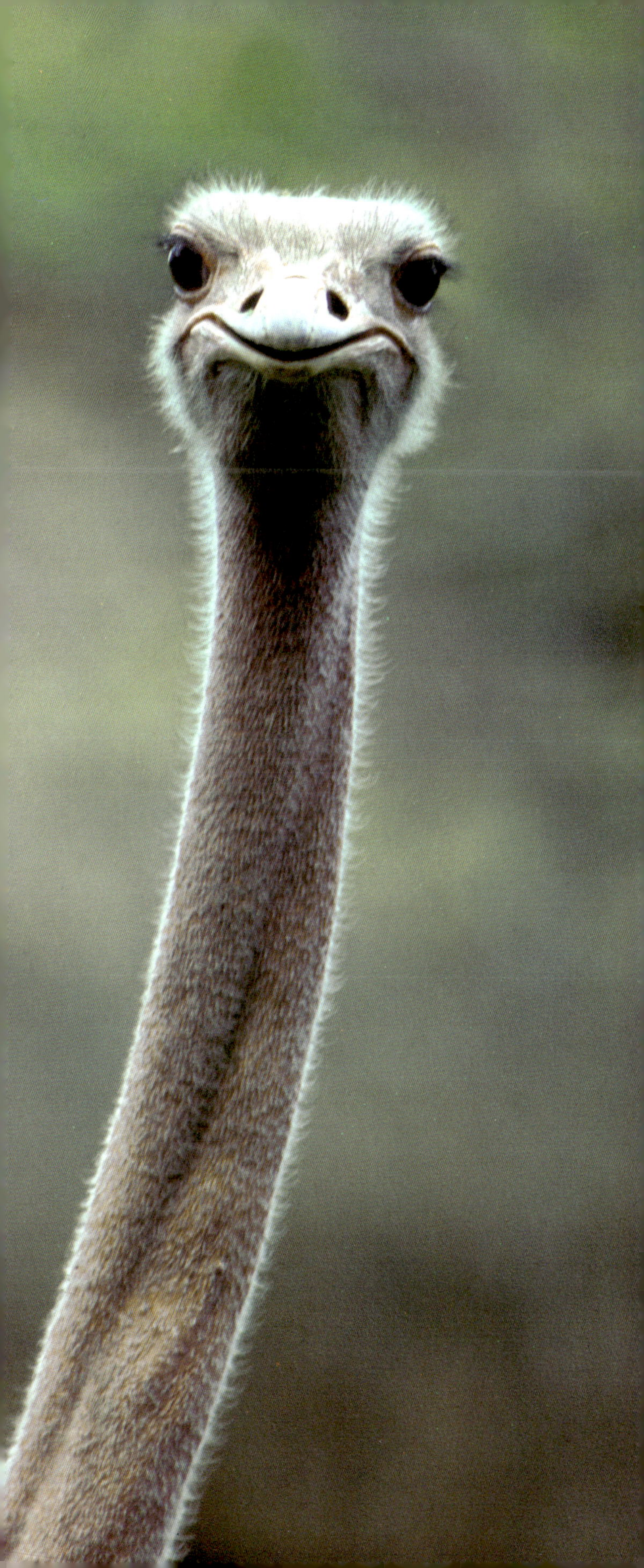

and can often be seen hanging by one arm while they use the other to pluck food from the trees.

The pelican's most striking feature is the large pouch under its bill, which it uses to catch fish. Before swallowing, the pelican must return to the surface and drain the water from its pouch.

↑ WHITE PELICAN

FLAMINGO POND (B3)

With their long legs, strange beaks, and bright pink plumage, flamingos are among the most recognizable of birds. At the zoo and in the wild, they are usually found near shallow, brackish water—for example, salty lagoons near the sea, salty lagoons in Africa's Rift Valley, or even salty lagoons high in the Andes Mountains of South America. The reason for this is their method of getting food: filtering out algae, small shrimp, crabs, and other organisms from large quantities of water. The flamingo's beak contains an effective filtration mechanism, and the size of the food that can be eaten depends on how fine the filter is. Different flamingo species have different filters and so eat different foods. The result is that, in nature, several species of flamingos can share the same habitat without competing for food.

↑ FLAMINGOS

→ KING VULTURE

In the wild, flamingos get their striking pink color from substances called carotenoids in the food they eat. In captivity, most flamingos receive food supplements containing carotenoids to maintain their color. Without these supplements, pink flamingos would fade to white after just one molt.

BIRDS OF PREY (B3)

If you take a look at an eagle, a hawk, or an owl, it's easy to see that these birds are hunters. Many birds of prey, or raptors, share certain features that help them capture their prey. These include hooked beaks for tearing flesh, extraordinarily acute vision, and, in ea-

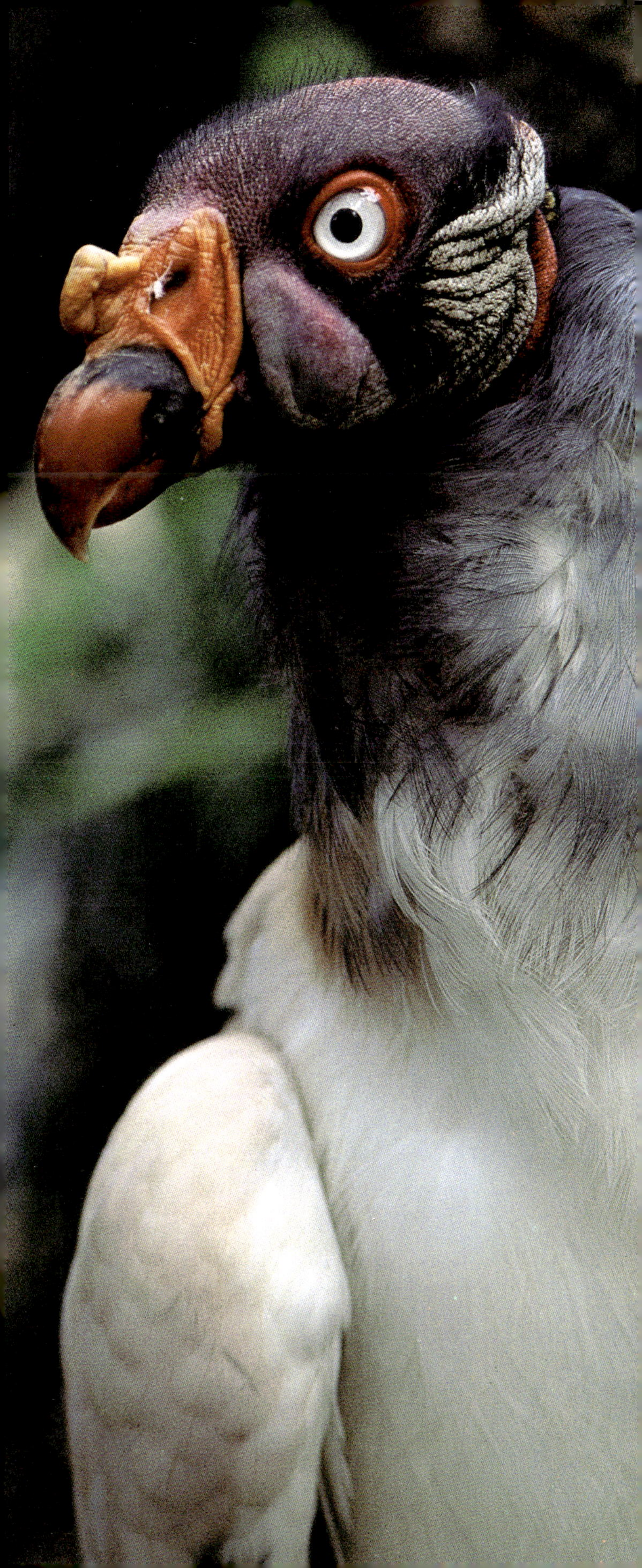

gles and owls, powerful grasping talons. For their size, birds of prey have disproportionately large eyes. Most also have a very high density of cones (the sensitive visual cells at the back of the eye) and two foveae (centers of focus), allowing them to see clearly both forward and sideways. (People have only one fovea, giving them clear forward vision.) Owls, which are most active at night, have a high density of rods (light-sensitive cells) for better vision in dim light. As a result of these visual modifications, many birds of prey can soar several hundred feet in the air and still spot a tiny mouse or squirrel scampering across the ground.

Although they are not hunters in the true sense, condors and vultures are classified as birds of prey. These birds do not capture live prey but rather feed on the carcasses of animals that have died naturally or been killed by other predators.

In their breeding season, North American bald eagles and several other raptor species perform spectacular "sky dances" as a prelude to mating, which takes place only after a nest has been built. Bald eagles build their nests, called eyries, in the topmost branches of tall trees. As more and more of our country's old growth forests are cut for timber, bald eagles find fewer suitable nest trees each season. Only by protecting and preserving raptor habitat and carefully monitoring the use of pesticides will we insure a future for bald eagles and other birds of prey around the world.

↑ BALD EAGLE

WILD HORSES AND GUANACOS (A3)

In a large pasture along the zoo's northwestern border, you can view the only truly wild horses still in existence: the Mongolian, or Przewalski's, wild horses. The Bronx Zoo was the first American zoo to exhibit these animals, now completely extinct in nature. These wild horses are easily distinguished from the "wild horses" of the American West (which are actually descendants of strayed domestic horses) by the mane. All wild members of the Equid family, including zebras, wild asses, and Mongolian wild horses, have erect manes. Domestic and feral horses have long manes that fall to one side.

↑ MONGOLIAN WILD HORSES

The guanaco, a New World relative of the camel, once roamed the South American plains in large numbers. In the 1500's, there may have been several million guanacos on the plains of Patagonia alone. Today, guanacos are threatened in the wild as a result of hunting for skins and because they are considered competitors of domestic sheep for grazing. Guanacos adapt to many different habitats, including deserts and plains, and are found at elevations up to 14,000 feet. They can go almost indefinitely without drinking, getting all the water they need from their diet of grasses and other plants.

↑ GUANACO

At the Children's Zoo, kids enter the homes of prairie dogs, spiders, and otters or test their jumping prowess against a bullfrog's.

CHILDREN'S ZOO (B4)

Helping children to see the similarities and differences between humans and other animals is one of the main goals of the Children's Zoo, which was completely rebuilt in 1981 to emphasize ecology, habitat, and interactive exhibits. Children and adults are invited to become Zoo Explorers as they investigate animal homes, locomotion, defenses, and senses.

Animals need homes for many of the same reasons people do: to raise their young, escape from bad weather, store food, and just rest and relax. And, like people, animals build different kinds of homes depending on where they live and what mate-

↑ DONKEY

↓ PRAIRIE DOG EXHIBIT

27

rials are available. Prairie dogs, which live on the grassy plains of western North America, dig extensive burrows underground. Composed of many different "rooms" linked together by tunnels, the burrows are large enough to house an extended family of these industrious rodents.

Other animals that build their own homes include night herons, which carefully construct their nests out of sticks, and spiders, which spin their web homes out of silk. Otters, on the other hand, look for a ready-made shelter, usually in a hollow log or beneath a tree stump.

There are many different ways of getting around, and sometimes very different animals move in surprisingly similar ways. Both bullfrogs and wallabies use their powerful hind legs to jump great distances; turtles and ducks use their webbed feet to propel them through the water; and both fish and alligators swim by sweeping their tails back and forth.

To survive in the wild, all animals need ways of defending themselves. Some take an active approach, arming themselves with "weapons" such as quills (the porcupine), horrible smells (the skunk), or even poison (the marine toad). Other animals try to hide from predators by blending in with their surroundings. The great horned owl, whose colors resemble those of a tree trunk, and the walking stick, which looks like a twig, are examples of animals that rely on camouflage. Still other animals, like the lizard, escape danger by running away. Outrunning predators isn't an option for slow animals like the tortoise, which stays put when danger threatens, taking refuge inside its hard shell.

Animals use many different senses to find out about the world. One look at the oversized ears of the fennec fox will tell you that this desert hunter relies on its hearing to find food and listen for danger. Animals that live in dark underground burrows, like the tarantula and the kangaroo rat, find out about their environment by touching it — the kangaroo rat feels its way through the subterranean tunnels with its whiskers, and the tarantula uses sensitive hairs on its legs to touch and identify objects. Some animals, like the tortoise, use smell and taste to explore their environments. And when snakes flick their long forked tongues, they're really "tasting" the air to find out if food or danger is nearby.

After you've learned something about how wild animals live, see and touch some

↑ FENNEC FOX
→ WALLABY

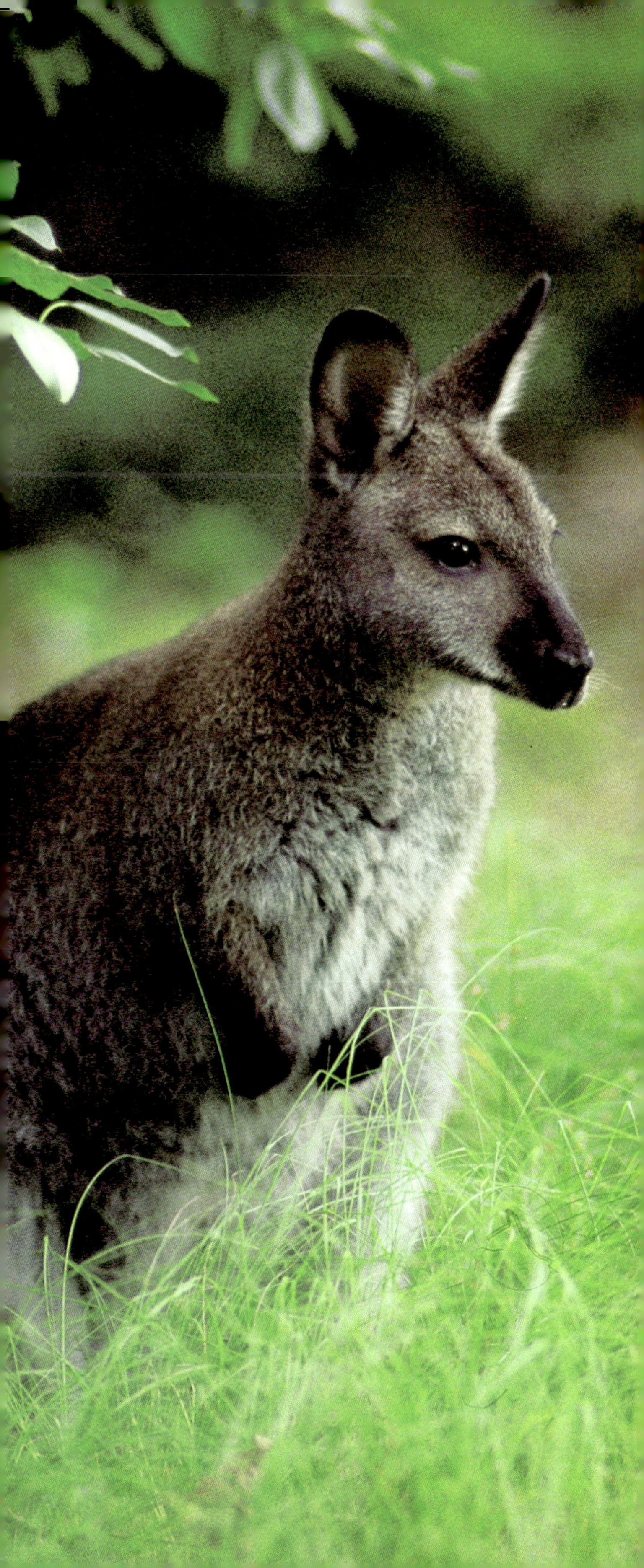

more familiar species in the domestic animal area. The reason we can touch and hold many domestic animals is that, unlike wild animals, they have been selectively bred for tameness. Thousands of years ago, humans began to tame and domesticate certain species—for the food or fiber they provided (chickens, pigs, cows, and sheep, for example), for companionship and help with hunting (dogs and cats), or just for their beauty (goldfish, cockatiels, and parakeets). Over many generations, some of these animals lost the ability to survive on their own and now depend on people to care for them.

In the domestic animal area you'll also find the Children's Theater, where members of the Zoo Education Department give demonstrations with live animals on a regular basis. There's also a photo booth, where you and your child can have your picture taken to commemorate your visit to the Children's Zoo.

> World of Reptiles houses a Conservation Station, where visitors learn about rare reptiles in the wild and see a reptile "nursery."

WORLD OF REPTILES (C6)

Some of the earth's most amazing animals are on view in the World of Reptiles. This imposing building, which opened in 1899, houses giant snakes, enormous crocodiles, turtles, deadly venomous serpents, and lizards. Bizarre amphibians, such as frogs and salamanders, are also on display.

One of the largest inhabitants of the World of Reptiles is a 24-foot-long reticulated python. Like boa constrictors, pythons kill their prey by constriction, coiling their powerful bodies around the prey and squeezing until the animal can no longer breathe. One of the differences between boas and pythons is that boas give birth to live young, while pythons lay eggs. A female reticulated python lays as many as 100 eggs, which she then warms by curling her body around them until they hatch. A newly hatched reticulated python measures about two feet in length, but it may eventually reach a length of more than thirty feet.

↑ AZURE POISON DART FROG

At the west end of the World of Reptiles, you'll see Chinese alligators basking on a muddy stream bank. The Chinese alligator's original

Yangtse River habitat has been lost as wetlands have been developed for human use. As a result, this impressive reptile is extremely endangered in the wild. Today, small numbers live in rice paddies and reservoirs.

The radiated tortoise, endemic to Madagascar, is one of the most brightly colored tortoises, named for the yellow-and-black sunburst pattern that decorates its back. Tortoises are known for their long lifespans—a few may reach 100 years of age or even older. The record may be held by a radiated tortoise named Malila Tui, which was presented by Captain Cook to the Queen of Tonga in 1776. It died in 1966 at the age of 189.

↑ RADIATED TORTOISE

↑ BROAD-SNOUTED CAIMAN

Radiated tortoises are endangered in the wild as a result of market hunting and illegal capture for the pet trade. A Malagasy native who collects a tortoise from the forest may sell it for a few dollars. But after that same tortoise has been smuggled overseas and sold to a collector, it may bring as much as $10,000. Illegal capture and smuggling of reptiles, whether for the pet trade or for sale of their skins, threatens many of the species you'll see in the World of Reptiles.

MOUSE HOUSE (B6)

Originally built in 1904 as the Small Mammals House, the Mouse House exhibit now focuses on some of the world's more than 1,700 rodent species. The building's 32 glass-fronted exhibits offer close-up views of the animals and feature naturalistic settings from four ecological zones: grasslands, tropical forest, woodland, and desert.

When people think of rodents, it's usually mice and rats that come to mind. But the order Rodentia includes a much wider range of animals than you might expect. Squirrels, beavers, prairie dogs, chipmunks, gophers, hamsters, gerbils, muskrats, woodchucks, chinchillas, and porcupines are rodents. Mouse House is home to the smaller members of the rodent clan.

Taxonomy Hall, a series of four exhibits, explains the difference among three suborders of rodents: squirrel-like, cavy-like, and mouse-like. The southern flying squirrel belongs in the first group. This tiny rodent has a thin membrane of skin connecting the front legs to the hind legs. By extending its four legs, the flying squirrel creates a kind of sail that enables it to glide through the air from tree to tree. Southern flying squirrels are found throughout the central and eastern United States. Unlike the familiar gray squirrel, flying squirrels are active only at night, so you may have some living in your yard and not even know it.

↑ PREVOST'S SQUIRREL

↑ ZEBRA MOUSE

The chinchilla, a small South American rodent, is prized for its soft fur. Hunting has almost completely eradicated chinchillas in the wild, but they are raised on fur ranches worldwide.

The mouse-like group is the largest of the three, containing more than 1,100 species. Named for the hard, spiny guard hairs that cover its coat, the Egyptian spiny mouse looks and acts quite a bit like the common house mouse. In the areas where it is found, North Africa and the Middle East, the spiny mouse competes with the house mouse for food and territory. The house mouse, however, is found on every continent around the world. In fact, the only rodent as successful as the house mouse is the equally adaptable Norway rat, which is also found just about everywhere on earth.

JOHN PIERREPONT WILDFOWL MARSH (B6)

One of the most spectacular swans you're likely to see can be found in the wildfowl marsh. Like other swan species, the black-necked swan has a snowy white body, but its long neck is a dark, glossy black, and there is a bright red knob at the base of the bill. Black-necked swans are native to South America. They build their nests among reeds in shallow water or on small islands in lakes or marshes. The young swans, called cygnets, are carried on the backs of the adults until they are quite large.

Another very recognizable marsh bird is the mallard. The male has a glossy green head,

↑ BLACK-NECKED SWAN

↑ MALLARD

thin white neck band, purple breast, and bright orange legs, while the female is a dull, speckled brown. The male's bright colors help him to compete for the female's attention during breeding season. By contrast, the female's job is to rear the ducklings, so it pays to be as inconspicuous as possible. This helps to explain the female's dull colors, which allow her to blend in with her surroundings so that she and her ducklings do not draw the notice of predators.

> Visitors to Congo Gorilla Forest can contribute their entrance fee to help save gorillas, elephants, and other animals in the wild.

CONGO GORILLA FOREST (B7)

The zoo's newest and most unusual exhibit, the Congo Gorilla Forest, is a representation of parts of the world's second largest rain forest, which stretches hundreds of miles across central Africa. The zoo exhibit encompasses 6.5 acres and provides a habitat for gorillas, guenons, okapis, mandrills, many other mammals, birds, reptiles, amphibians, fishes, and invertebrates, as well as more than 15,000 plants of 400 different species.

Entering the Congo Forest, you'll see strikingly marked black-and-white colobus monkeys living in the trees. Next, you'll pass through a huge fallen log, where you may glimpse the rare okapi, a secretive and solitary animal that looks like a cross between a zebra and a giraffe. Okapis are closely related to giraffes and have many giraffe-like characteristics, including skin-covered horns and a long black tongue that is used to gather leaves into the mouth. However, giraffes are found on Africa's open savannas, while okapis are rain forest dwellers.

Okapis are found only in the rain forests of northern Zaire, where they are still hunted for their meat, which is highly prized. Wildlife Conservation Society scientists have worked to learn more about okapis in the wild for many years, and WCS was instrumental in establishing the 5,000-square-kilometer Okapi Reserve in Zaire's Ituri Forest. Near the

↑ OKAPI
↑ LONG-TOED RUNNING FROG
PAGE 34: COLOBUS MONKEYS
PAGE 35: MANDRILL

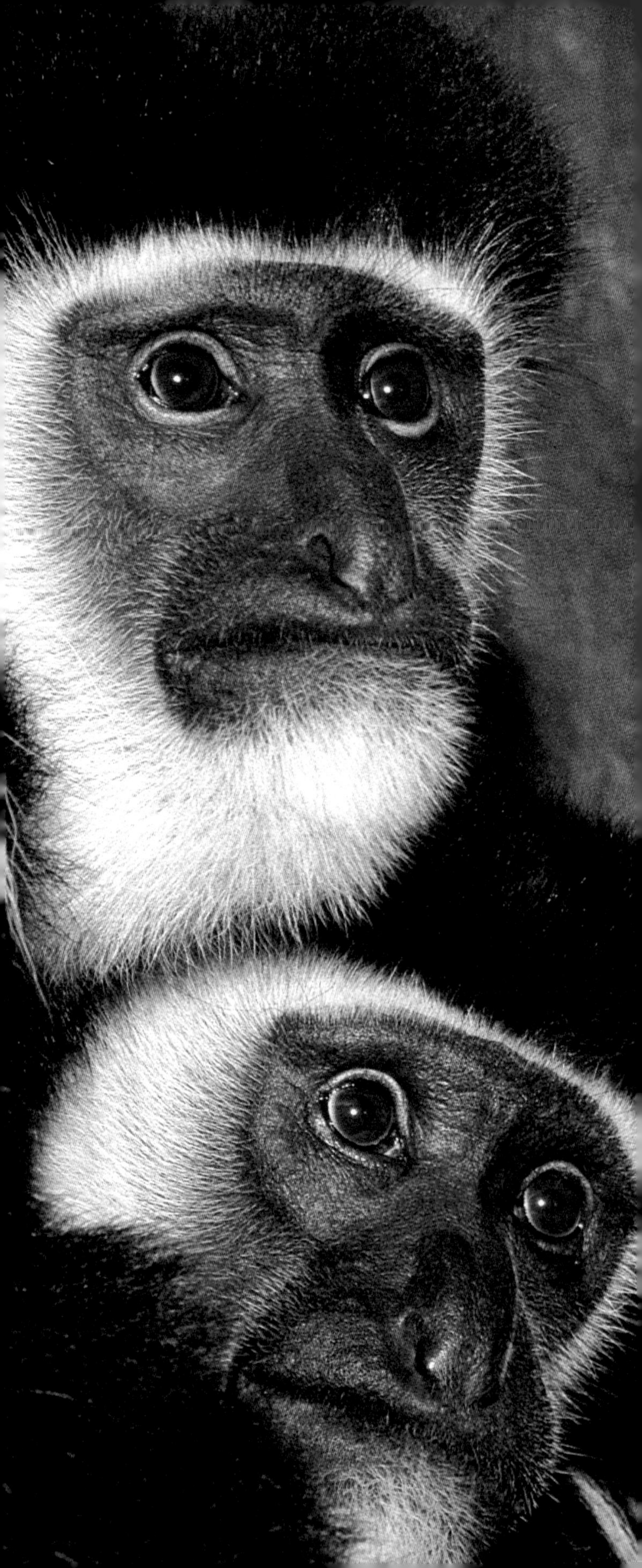

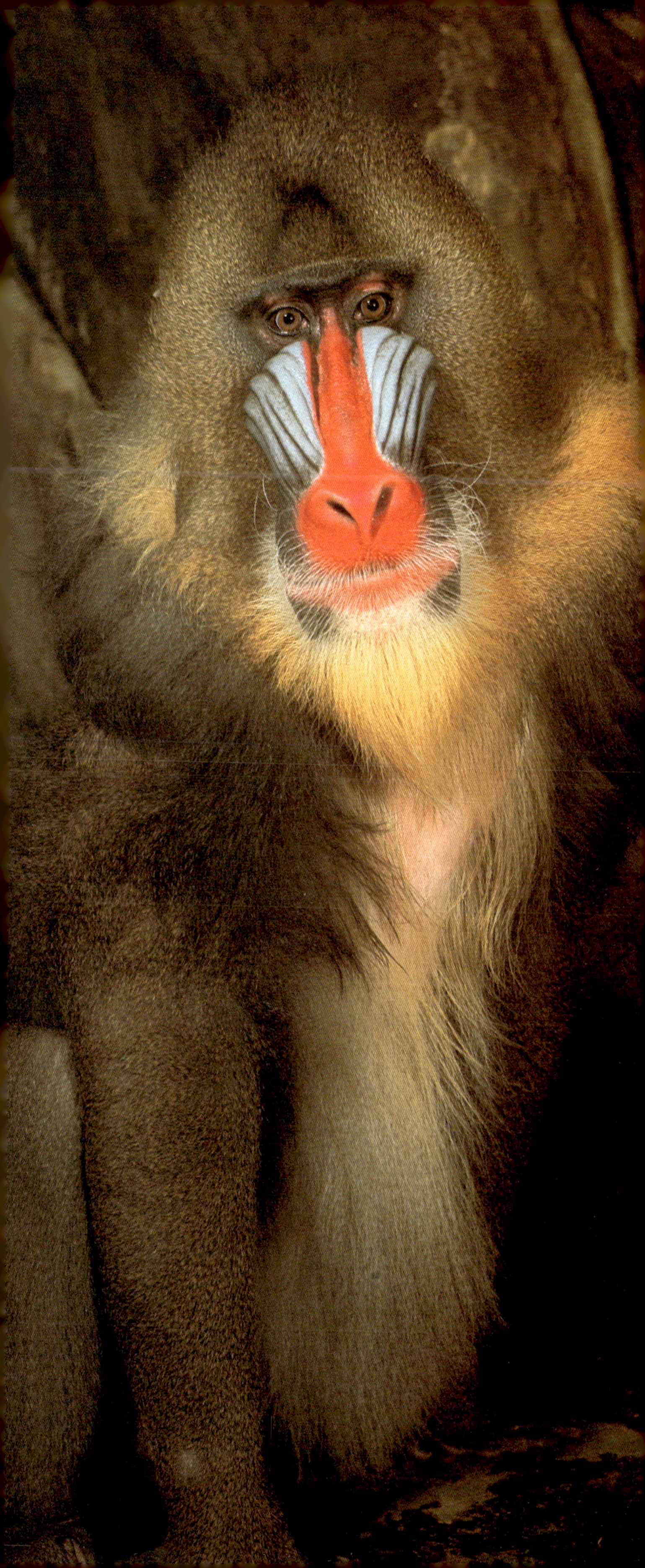

okapi exhibit, you'll be able to see a "field station" much like the site used by WCS researchers Terese and John Hart in the Ituri Forest, and you'll get an idea of some of the equipment and procedures that field scientists use to study animals and their habitats.

After leaving the okapis, you'll continue along the Rain Forest Trail through a jungle of lush tropical plants. Animal sounds recorded in the rain forests of Africa will make you feel as though you have been transported to the Congo region, and clues and queries along the trail will challenge you to spot wildlife and solve problems much as a WCS field scientist would do.

Your next stop is the Living Treasures of the Rain Forest Gallery. As you pass a large fallen tree, be on the lookout for mandrills and DeBrazza's monkeys in the trees and red river hogs along the nearly dry riverbed. The mandrill is a huge forest monkey and is the most brilliantly colored of all mammals. Mandrills inhabit only the dense rain forests of west central Africa. Like many other primates, mandrills live in harems consisting of one male, several females, and their young. Since daughters tend to stay with their mothers as long as they live, the females in a group are usually related to one another. Males, on the other hand, usually leave the group when they reach adolescence.

The male mandrill's coloration—a bright red nose and muzzle flanked by patches of purple-blue, and a bare blue-and-mauve patch on the rump—are brightest in dominant males and may act as signals to troop members in the thick rain forest vegetation.

Much smaller than the mandrills, the DeBrazza's monkey, a member of the guenon family, is also an inhabitant of the central African rain forest. Many guenon species are known for their beautiful coat colors and patterns, which are sometimes striped or speckled, and for their whiskers and prominent tufts of facial hair. Male guenons may have more brightly colored coats and are usually larger than females. Near the monkeys, you may also see red river hogs foraging along the edge of a stream. These beautiful wild pigs have been eradicated throughout much of their range because of the "bush meat" trade. In addition to the animal exhibits, this gallery features several interactive displays, where you'll get a chance to explore thermal imaging (how some animals sense the presence of others by temperature) and related specialized

↑ WOLF'S GUENON

→ DEBRAZZA'S MONKEY

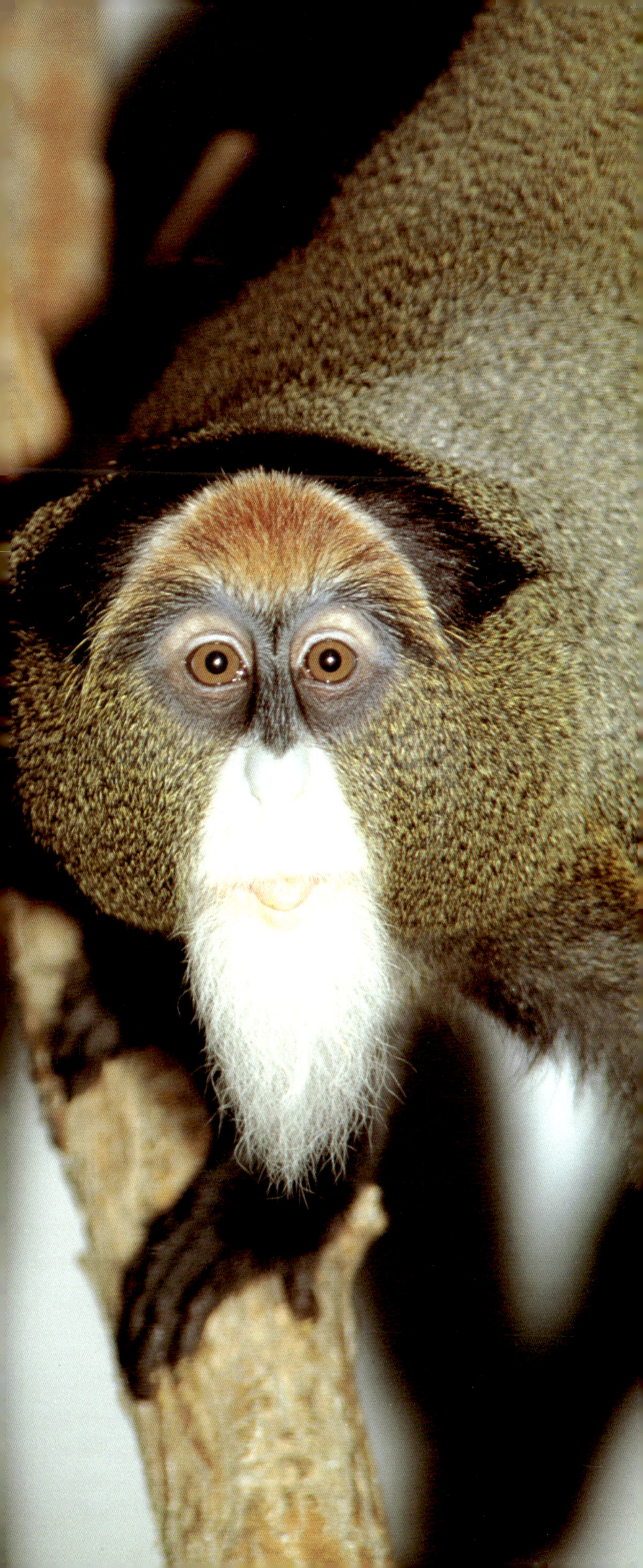

↑ RED RIVER HOG

→ LOWLAND GORILLAS

PAGE 40: MALE GORILLA

PAGE 41: BABY GORILLA

adaptations of rain forest animals.

Thus far, all the animals you've seen in the Congo Forest have been large and easy to spot. In the wild, however, there is a wealth of wildlife that is often barely noticeable. These creatures, which make up the vast majority of rain forest life, include fish, frogs, snakes, lizards, turtles, and countless species of insects. In fact, the world's rain forests are repositories of a greater variety of animal and plant species than any other habitat on earth. Some of the lesser known but nonetheless fascinating animals you'll see are elephant-nosed mormyrids (fish that navigate and communicate using electricity), upside-down catfish, dwarf crocodiles, goliath beetles, and red-eyed assassin bugs. The survival of these species, as well as those yet to be discovered, is one of the reasons for protecting and preserving the earth's rain forests.

Your next stop is the Conservation Showcase, where you'll see a huge tree half felled by a chain saw. This dramatic scene is an all too real reminder of the uncontrolled logging that continues to threaten the forests and wildlife of central Africa. Surrounding exhibits address conservation dilemmas, such as deforestation, overpopulation, and natural resource depletion. You'll also see how WCS is working to conserve these vitally important forests in ways that benefit both wildlife and local people.

Next, you'll enter the Conservation Theater, where a brief video will transport you to central Africa to visit WCS scientists at work. The focus of the program is the Society's ongoing work in gorilla conservation, which began in 1959 when WCS Director for Science Dr. George Schaller conducted his pioneering studies of mountain gorillas in Rwanda and Zaire. Since then, WCS has supported numerous gorilla conservation projects, including behavior studies by Dian Fossey and work by WCS scientists Drs. Amy Vedder, Bill Weber, Mike Fay, Lee White, Jeff Hall, and others.

As the video program concludes, the screen slides up and curtains open to reveal a troop of gorillas in a large outdoor habitat. You will then move into the Great Apes Gallery, the heart of the Congo Gorilla Forest. Tall glass windows provide a panoramic view of the zoo's two western lowland gorilla troops, one of the largest breeding populations of zoo gorillas in the world.

Gorillas are the largest of the world's living primates and, along

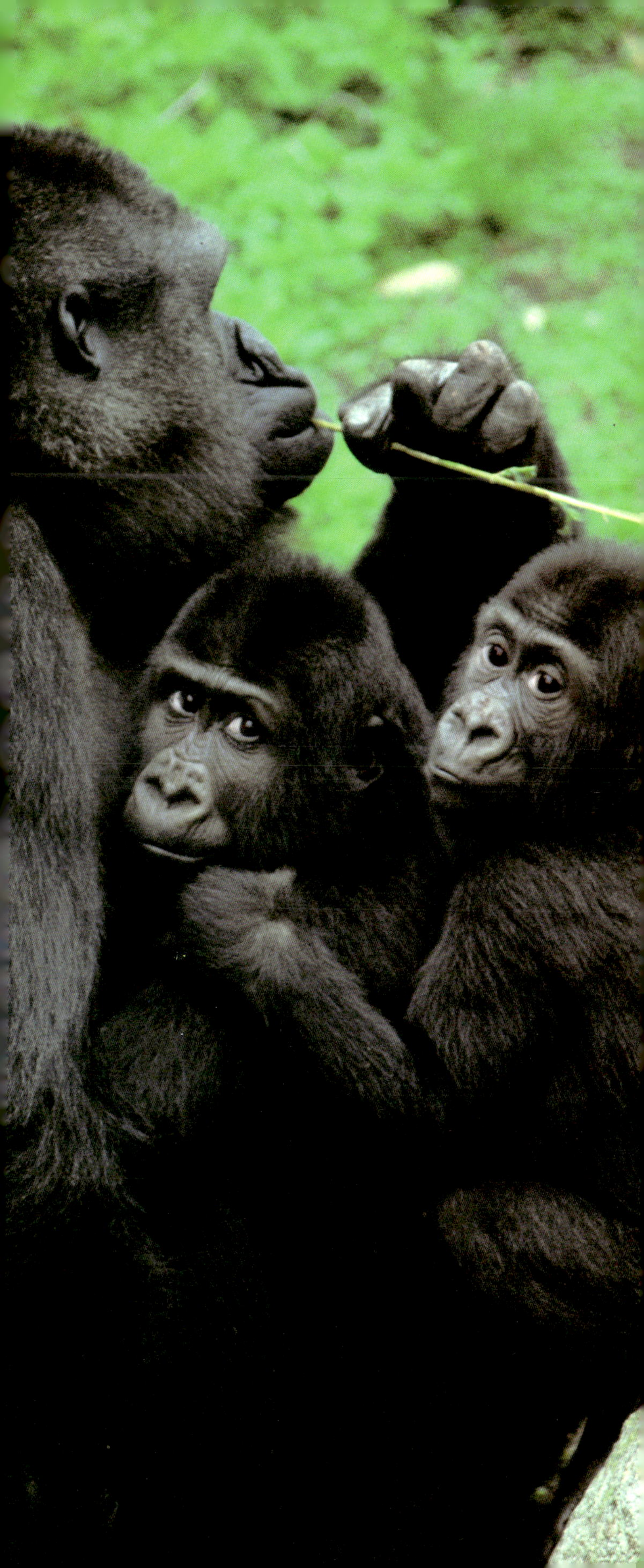

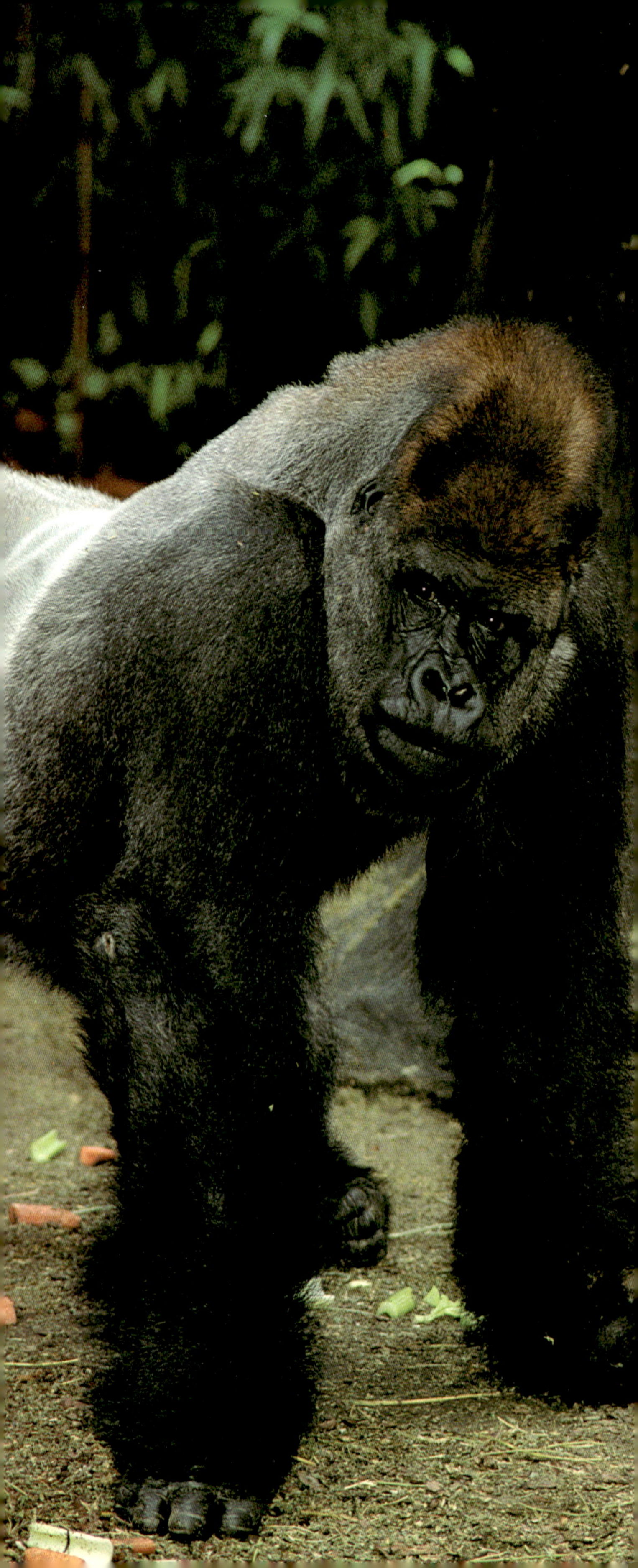

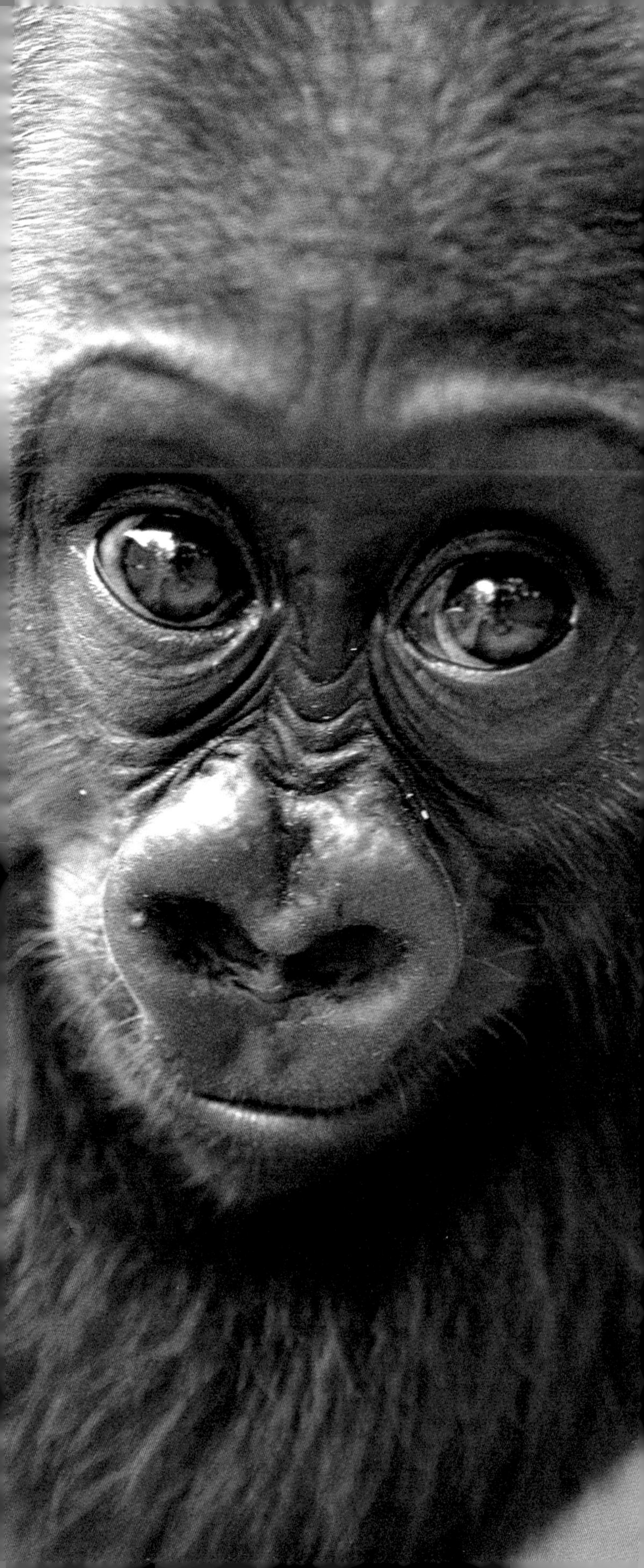

with chimpanzees and bonobos, are the apes most closely related to humans. In nature, gorillas live in groups consisting of one dominant male (usually called the silverback because of the silvery white saddle of fur covering the back), several adult females, and their offspring. Gorilla groups are very stable, with the same individuals remaining together for months or even years at a time. In contrast to their "King Kong" image, gorillas are actually very peaceable. They feed almost entirely on plants and spend much of their time browsing for food on the forest floor. Before leaving the Great Apes Gallery, be sure to try some of the interactive exhibits that focus on gorilla anatomy and behavior and explore their evolutionary links with humans.

Western lowland gorillas are one of three races of gorillas (the others are Grauer's gorillas, also known as eastern lowland gorillas, and mountain gorillas) that are increasingly imperiled in the wild. Illegal hunting for food and destruction of their forest habitat are the most serious threats. You can find out more about WCS gorilla conservation projects in the Conservation Choices Pavilion, which is reached through a glass tunnel that passes directly through the gorillas' habitat. Here you will be able to review a series of actual Wildlife Conservation Society field projects in central Africa, and you'll decide which project will receive your entry donation, making your visit to the Congo Gorilla Forest an act of conservation in itself.

> Baboon Reserve includes special "field stations" for a close encounter with gelada baboons of Ethiopia.

BABOON RESERVE AT SOMBA VILLAGE (C6–7)

Designed to resemble the grassy highlands of Ethiopia, the treeless hillside of the Baboon Reserve is home to three troops of gelada baboons, a species that is becoming increasingly rare in nature. The only true grazing primate, the gelada lives almost exclusively on grasses. Because of its high degree of diet specialization, this primate cannot easily adapt to different habitats. Unfortunately, much of the land on which the gelada depends is being converted into farmland, and gelada populations are dwindling. Hunting is an additional threat: in the south-

↑ GELADA BABOONS

→ GELADA BABOON

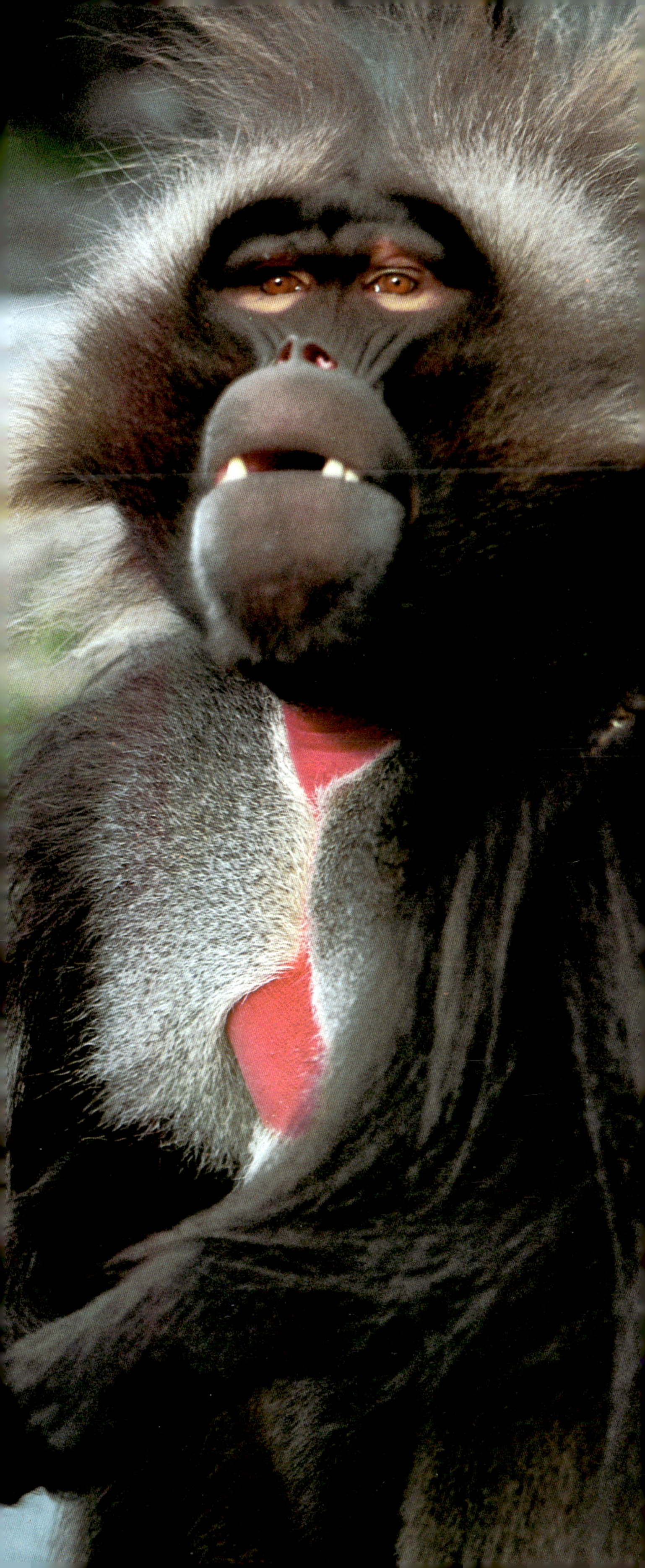

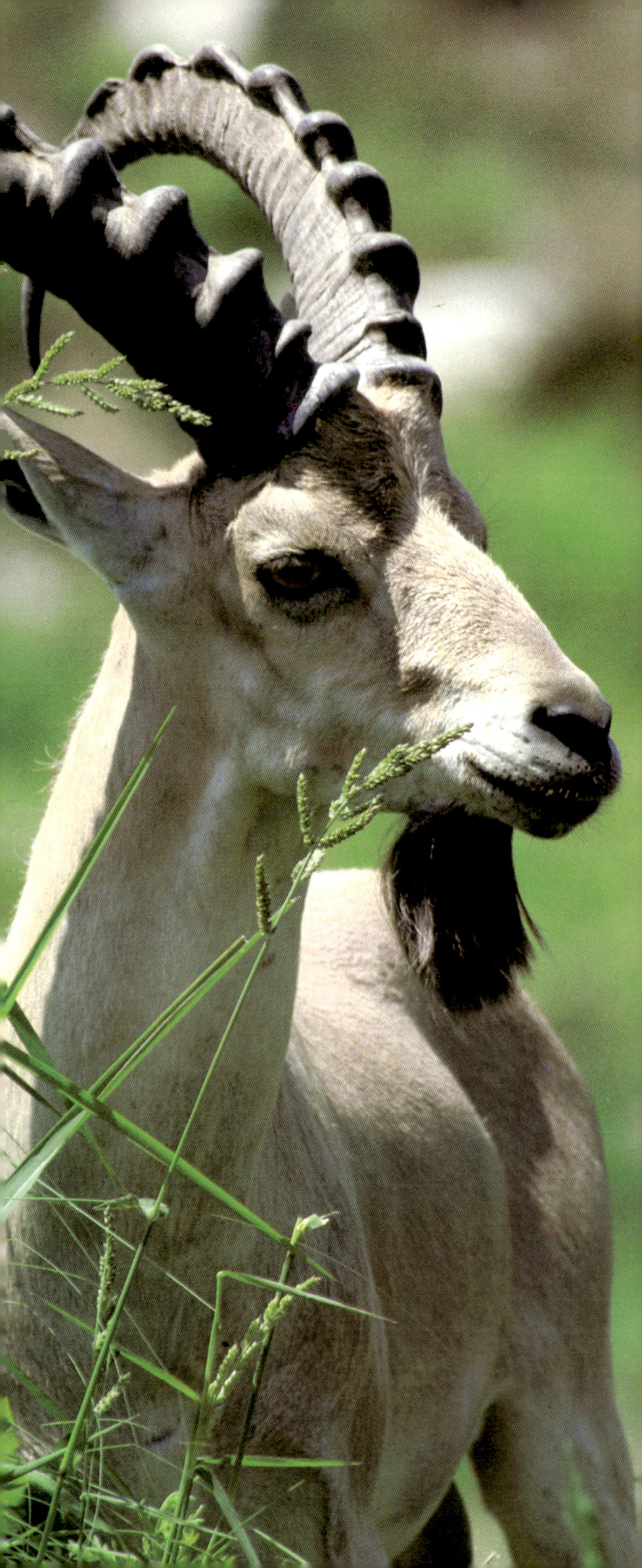

ern part of their range, male geladas are hunted intensively for their luxuriant manes, which are used by the local people in headdresses and capes.

Geladas live in bands consisting of 35 to 350 individuals. The bands are made up of several smaller troops, either all-male groups, or harems, made up of one male, several females, and their young. Although one might expect that the dominant male would keep the harem together, it actually appears that social bonds among the females are most important for group cohesiveness. In the wild, when grazing conditions are good, several troops may come together to form temporary herds of as many as 700 geladas.

The gelada baboons can be seen both from the Somba Village and from "field stations" in the reserve. If you watch carefully, you may be able to see not only social interactions among troop members but also encounters among the different troops. Geladas rely on sounds and visual signals to communicate with one another, and scientists have identified approximately 25 different calls, each of which has a different meaning. One of the most noticeable visual signals used in gelada communication involves the hairless pink patch of skin on the chest. It is thought that color

↑ GELADA BABOON BABY
← NUBIAN IBEX

changes in this area signal sexual receptivity. In females that are ready to mate, this area gets darker in color and becomes covered with raised white spots—a change that is virtually impossible for other geladas to miss.

The gelada baboons share their habitat with the Nubian ibex, a wild goat found in northeastern Africa and the Arabian regions. Like other mountain goats, Nubian ibex are expert climbers and live on mountain slopes at elevations of up to 20,000 feet above sea level. Ibex herds move seasonally—up into the high mountains during the summer and down to lower levels where vegetation can still be found in the winter.

In the wild, ibex herds are segregated by sex. Herds consist either of females and young or "bachelor" groups of males. In all-male groups, like the one at the Bronx Zoo, it is not uncommon to see ritualized

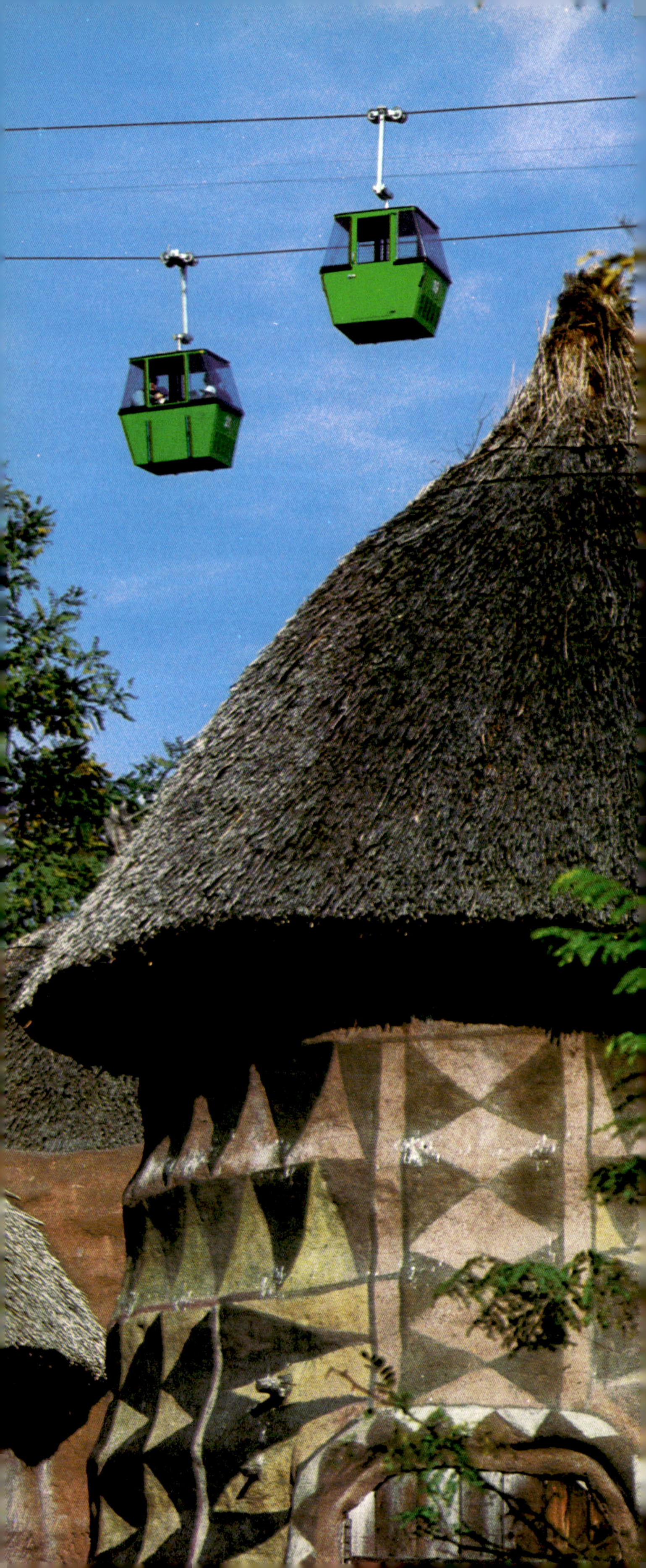

fighting to establish dominance. Watch closely and you may witness two opponents rearing up on their hind legs and then crashing into each other with their long, curving horns. Although quite spectacular, these fights rarely result in death or even injury. Combat reaches a peak during the breeding season, when males join the female herds and compete for mates. The males and females often remain together for the winter and early spring, but by the time the kids are born, in May or June, the males have already departed to resume their bachelor existence.

Like the gelada baboons, Nubian ibex are threatened by loss of habitat and hunting. The Wildlife Conservation Society has sponsored several field programs in Ethiopia designed to safeguard the habitat of vanishing species.

After you've watched the wildlife, follow the pathway through sandstone cliffs to the Fossil Dig. Here you'll see a re-creation of an excavation site, including fossils of geladas, early humans, and other species, as well as some of the tools used by scientists to uncover these clues to the past.

At the southeast corner of the Baboon Reserve is the Somba Village, a grouping of circular thatch-roofed huts decorated with geometric painted patterns. In this little village, similar to those found in Cameroon or Togo, you'll find food and refreshment stands and a gift shop. Covered and terraced seating areas provide views of the market and across a stream to the rugged hillside of the Baboon Reserve.

↑ ROCK HYRAX
↑ FOSSIL DIG
← SKYFARI ABOVE SOMBA VILLAGE

Another exciting way to see the Baboon Reserve and Somba Village is from the Skyfari. Hop on at Skyfari West (near the Southern Boulevard pedestrian entrance) or Skyfari East (near the Asia parking lot) and get a bird's-eye view of the baboons and ibex directly below, Africa to the south, and the Big Bears and Himalayan Highlands to the north. Skyfari is open from April through October.

CARTER GIRAFFE BUILDING (C7)

Sometimes called "the animal built by a committee," the giraffe does indeed look as though it was put together out of spare parts. Its neck is disproportionately long compared to its body, and the hind legs are shorter than the front legs, so that the animal slopes continuously from head to tail. Giraffes look so awkward and defenseless that it's hard to imagine how they survive attacks by lions and other predators inhabiting the African plains. In fact, giraffes are formidable opponents. If attacked, they can deliver a powerful kick with their huge hooves or turn and run away from predators at speeds of nearly 40 miles per hour. And with their superior eyesight, smell, and hearing, giraffes have an "early warning system" that allows them to avoid most predators in the first place.

For the most part, giraffes are unassuming animals that spend more than half their waking hours browsing on leaves. Because of their great height, they feed on vegetation that cannot be used by other browsing animals or by domestic livestock. Because they do not compete with livestock, giraffes have been tolerated by cattle ranchers.

Near the giraffes, you'll find the fastest land animal in the world: the cheetah. When pursuing a gazelle or other animal, a cheetah can reach speeds of up to 60 miles per hour. The cheetah's slim body and flexible spine enable this cat to make amazingly long strides. And for added traction while running, the cheetah's claws are exposed rather than covered by a sheath.

Cheetahs inhabit the African plains along with many other carnivores, including lions, leopards, jackals, hyenas, and wild dogs. To survive, the cheetah must specialize in a particular habitat and style of hunting. Because it is so fast, the cheetah does well on open ground, where it can run down its prey. But like all sprinters, the cheetah has limited endurance—it can run at top

↑ BARINGO GIRAFFE BABY
↑ CHEETAHS
→ BARINGO GIRAFFE

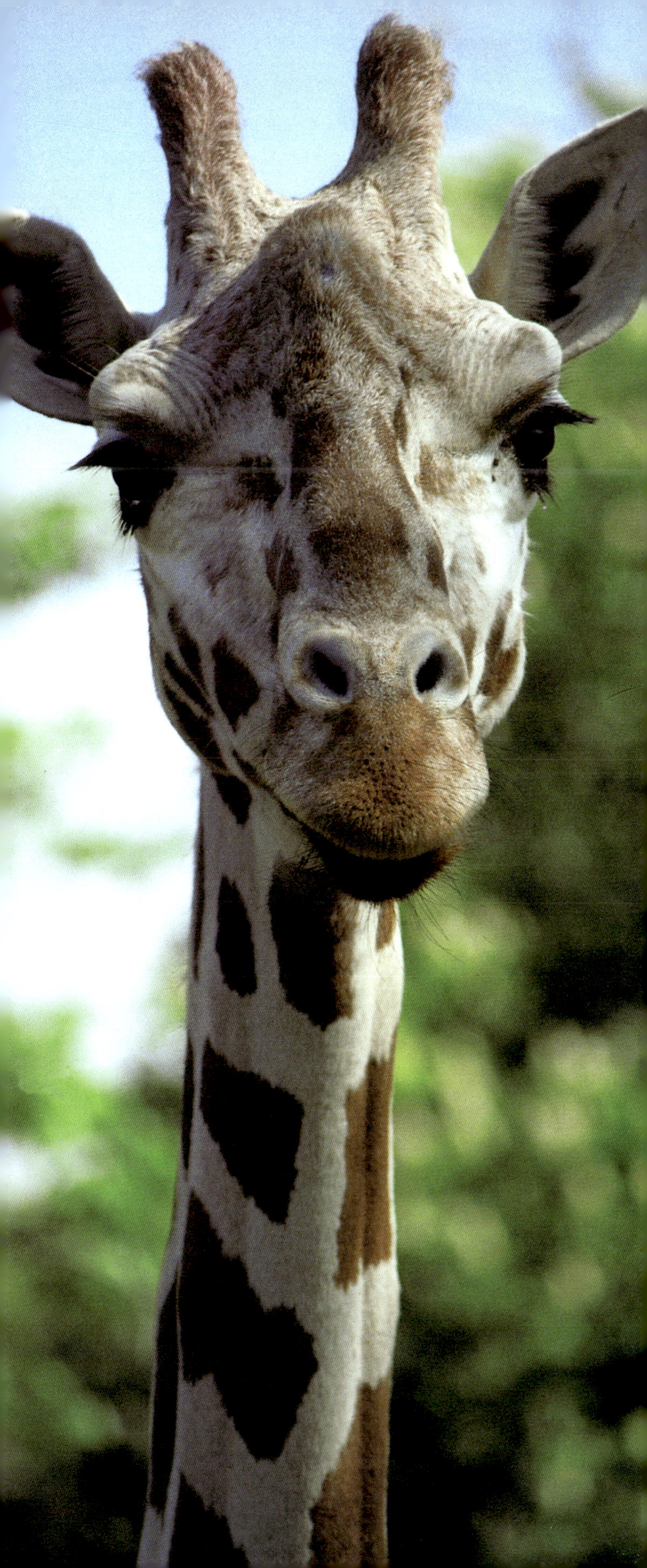

speed for only about 30 seconds. Because of this, its preferred habitat has enough vegetation so that the cheetah can sneak up to within a few hundred feet of its prey before giving chase. In addition to inhabiting a slightly different habitat, cheetahs minimize competition with other predators by hunting during the day, when most lions and leopards are napping.

Cheetahs are among the most endangered large cats. The most serious threats are poaching for their spectacular spotted coats, human encroachment into the open land the cheetah needs to survive, and shooting by farmers to protect their herds of goats and sheep.

AFRICA (C7)

In this re-creation of a typical African plain, a pride of lions appears to share the grassy plain with their typical prey: zebras, nyalas, Thomson's gazelles, and blesbok—thanks to concealed moats that separate the hunters and the hunted.

Lions, the most social members of the cat family, live and hunt in groups called prides. In an average pride, which consists of 1 to 6 males, 4 to 12 females, and their young, there is a strict division of labor. The job of the males is to defend the pride's territory and the females against marauding male lions, while the females are in charge of hunting. Each sex is perfectly suited to its task: the male's large size and impressive mane make him a formidable enemy, but these characteristics also make him very conspicuous. The smaller, less noticeable females are not robust enough to defend the territory but are able to blend in with their surroundings as they stalk their prey. Because they hunt in groups, lions are able to bring down prey animals much larger than themselves.

In the wild, zebras often fall prey to lions and other predators, and just one look at a zebra will tell you why. With its kaleidoscopic black and white stripes, the zebra is one animal that can't hide from predators. Instead of camouflage, zebras rely on their keen eyesight and hearing to alert them to danger. Living in a large herd is also helpful—with hundreds of zebras on the lookout, it's almost certain that someone will spot an approaching lion or cheetah. And, ironically, the best place for a zebra to hide is in the middle of a group of other zebras. No one really knows why zebras

↑ MARIBOU STORK
→ LION

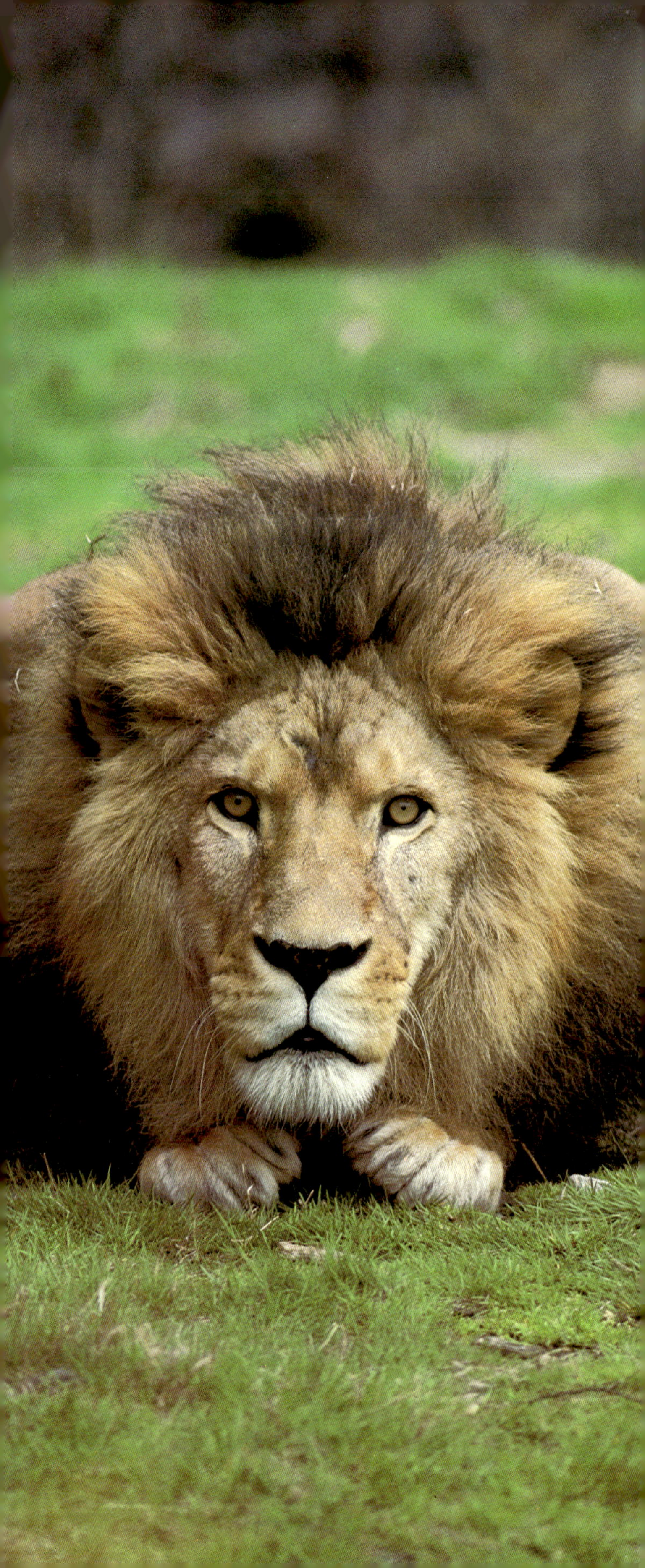

have stripes, but one of the most likely explanations is that it serves as a visual signal to keep the herd together.

Three other prey species found on the African plains are nyalas, Thomson's gazelles, and blesbok. Although they look quite different from one another, they are all members of the bovid family, which includes antelopes, cattle, bison, goats, and sheep. These three species, like all bovids, are browsing or grazing animals and spend most of their waking hours feeding on grasses or bushes. In the wild, these animals congregate in herds. As with zebras, living in a group offers nyalas, gazelles, and blesbok some protection against predators. Another defense is their running ability—Thomson's gazelles can reach speeds of up to 50 miles per hour.

↑ ZEBRA

↑ NYALA

Nyalas, blesbok, and Thomson's gazelles all have long, slender horns, but they are used mainly in encounters and displays among themselves rather than to defend against attackers.

> At JungleWorld, tour the steamy forests of Southeast Asia, home to tree kangaroos, proboscis monkeys, gibbons, leopards, and crocodiles.

JUNGLEWORLD (E9)

Occupying a 37,000-square-foot building near the Bronx Park South entrance is JungleWorld, a detailed re-creation of four typical Asian habitats: a dry scrub forest, a mangrove forest, where the twisted roots of mangrove trees grow into salt water, and the largest area, a 300-foot-long habitat that extends from lowland rain forest to montane rain forest.

At the JungleWorld entrance, you'll see a display of colorful Indian and Balinese banners and bells and giant Asian bamboos. You'll also learn some amazing statistics about tropical forests, including the fact that the world's tropical forests cover only 6% of its land area but are home to over

↑ TREE KANGAROO

→ GHARIAL

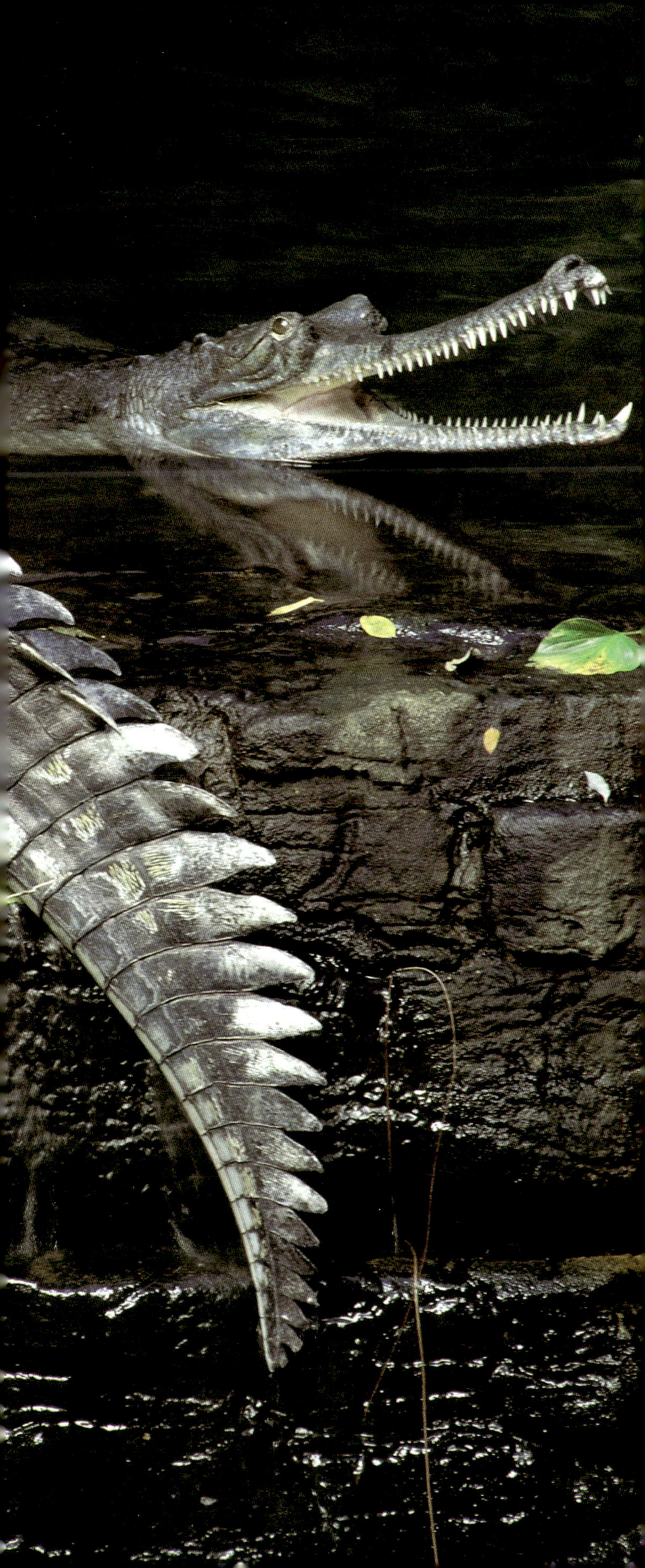

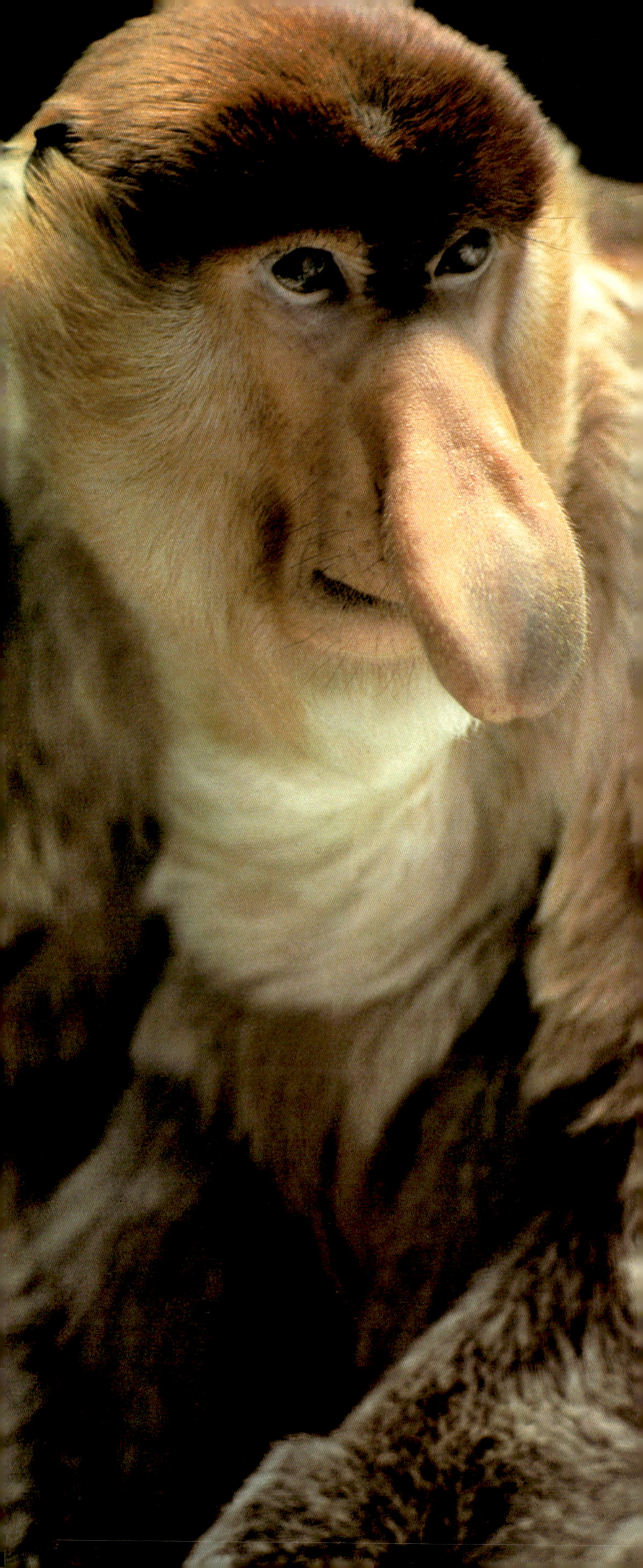

50% of all living species of plants and animals. After leaving the entrance hall, you'll find yourself in an area designed to resemble a Southeast Asian scrub forest. Here you will find tree kangaroos using their long furred tails for balance as they move among the branches. Tree kangaroos are smaller than their ground-living relatives, and they lack the greatly enlarged hind limbs that make other kangaroos such good jumpers. Although mainly arboreal, they descend to the ground fairly frequently and get around by hopping, although not with the agility of the terrestrial species. One interesting thing you may notice about the tree kangaroos is that the thick fur on the back of the neck grows in a reverse direction. Since these animals usually rest with their heads lower than their shoulders, the "backwards" fur probably acts as a natural water-shedding device.

Another arboreal inhabitant of the scrub forest is the binturong, also called the bear cat. In fact, the binturong is neither a bear nor a cat but is a member of the civet family. Like other civets, to which it is closely related, the binturong produces a strong-smelling substance (referred to as civet or civet oil), which it uses for scent-marking. It is said that the binturong's scent is similar to cooked popcorn! Binturongs are active mainly at night, when they move through the trees in search of food. The binturong grasps branches with its muscular prehensile tail, while pulling fruits, leaves, and shoots into its mouth with its forepaws. Binturongs also eat small birds and have been reported to swim in rivers and catch fish.

↑ BINTURONG
← PROBOSCIS MONKEY

Moving into the Mangrove Forest, you'll encounter a troop of proboscis monkeys. One look at an adult male will tell you where this species got its name: from the large, pendulous nose. The fact that only males have this outstanding appendage suggests that the large nose may help attract females for breeding and alert other males to the presence of a male among the females.

Proboscis monkeys make their homes near fresh or brackish water and are often found in mangrove swamps, where they rely on the mangrove trees for feeding, resting, and sleeping. These monkeys are excellent swimmers, both on the surface and underwater. Most of their diet consists of leaves and occasionally fruits and flowers. In the wild, proboscis monkeys live in harem groups, consisting of one male, several females, and their young (which are born with bright blue facial skin). Because of clear-cutting of mangrove swamps and human disturbance, proboscis monkeys are becoming increasingly rare.

↑ OTTER

You may catch a glimpse of the world's smallest otter species, the Asian small-clawed otter, scampering along the muddy banks and among the roots of the mangrove trees. This little otter feeds on fishes, crabs, and mollusks, crushing the shells with its large, broad teeth. The small-clawed otter is more "hand oriented" than other otters and uses its sensitive forepaws to search for prey by touch alone. (The German name for this animal is "finger otter.") Unlike the solitary river otters of North America, the Asian small-clawed otter is quite social, living in extended family groups. After breeding, male and female otters stay together, and the male helps to raise the young, which remain with their parents until they are about a year old.

Past the Mangrove Forest is the Lowland Rain Forest exhibit, home to a large troop of silvered langurs. These arboreal, leaf-eating monkeys are just one of more than ten species of leaf monkeys found throughout Asia. Silvered langurs live in fairly large groups, consisting of one or a few males, a greater number of females, and their young. Like many other inhabitants of tropical rain forests, these monkeys are becoming very rare in the wild, due to hunting and destruction of their forest habitat.

↑ SILVERED LANGUR
→ WHITE-CHEEKED GIBBON

Even better adapted to a life in the treetops are the white-cheeked gibbons, ranging from lowland rain forest to montane rain forest. Almost exclusively arboreal, they use their long, powerful arms to swing from branch to branch, traveling up to 15 feet in a single swing. Gibbons are monogamous and live in family groups consisting of a male and female and their offspring. Gibbon families are highly territorial and engage in elaborate "songs" to advertise their presence and keep other gibbon groups away. The songs, which usually take the form of a duet between the male and female (with occasional accompaniment from their offspring) may also serve to strengthen the bond between the pair.

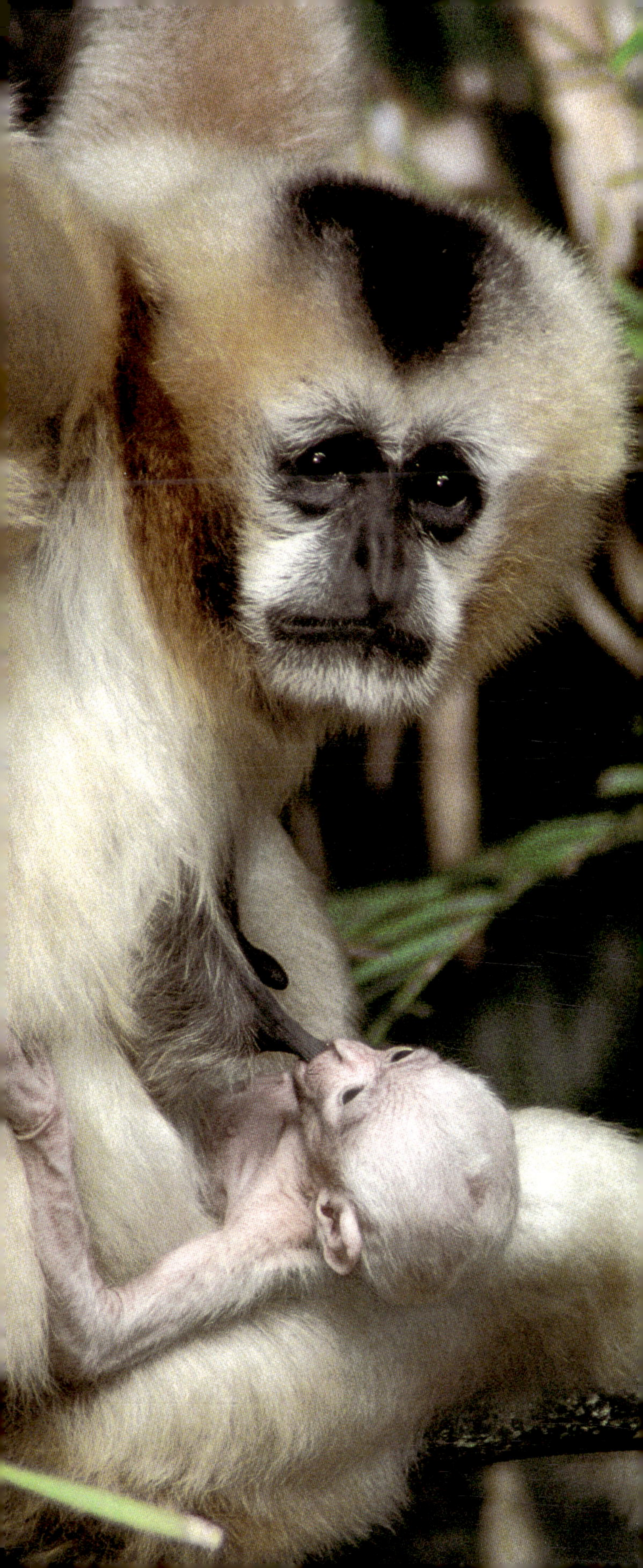

In the wild, gibbons, as well as silvered langurs, sometimes fall prey to leopards, which are also frequently found in the trees. The coat color of leopards varies, ranging from pale tan to brown to nearly black, and always covered with black spots. Black leopards were once thought to be a distinct species called the black panther. In fact, a "black panther" is simply a black-coated form of the leopard. If you look closely, you may be able to see the familiar black spots showing faintly against this cat's dark fur.

↑ BLACK LEOPARD

The leopard spends more of its time in the treetops than most other cats, either resting or surveying its surroundings for potential prey. A leopard may even drag its victim up into a tree, where it can savor its meal out of reach of scavengers.

On the beach or in the water near the leopard are the strikingly marked black-and-white Malayan tapirs. Tapirs are among the strangest large mammals in the world. Their somewhat porcine appearance and elongated nose sometimes lead people to believe that they are related to pigs or even to elephants. In fact, the horse and the rhinoceros are the tapir's closest relatives. In the wild, tapirs are threatened by habitat destruction and by hunting for food, sport, and their thick, leathery skins. You can visit more Malayan tapirs, along with Asian elephants and African black rhinos, in Zoo Center.

No less spectacular than the jungle's monkeys, leopards, and tree kangaroos are the smaller, often-overlooked creatures that actually

↓ ASIAN GREEN MANTID

↑ WALKING STICK

make up the bulk of the tropical forest's biomass. JungleWorld's largest gallery, The Unseen Multitude, will give you a close-up view of tiny forest inhabitants such as giant walking sticks, strange mantids, and curious lizards.

Deep pools in a river are home to the gharial, one of the most unusual-looking crocodilians you're likely to meet. The snout is extremely long and narrow and lined with sharp, needle-like teeth. Mature males have a swelling on the end of the snout that is said to resemble a type of Hindu earthenware pot called a *ghara*. Some authors claim (and it certainly seems likely) that this is the origin of the word gharial, the common Indian name for this reptile. Gharials can grow to lengths of more than 20 feet. They spend more time in the water than most other crocodilians and feed mainly on fish.

The highly endangered Rodrigues fruit bat, another tropical forest resident, is but one of nearly 1,000 species of bats that inhabit tropical and temperate regions the world over. (You may be surprised to learn that nearly one out of every four mammal species is a bat; bats and rodents together comprise two-thirds of all mammalian species!) As its name suggests, the Rodrigues fruit bat feeds chiefly on fruit. Because of this dietary specialization, fruit bats live only in the tropics, where ripening fruits can be found all year long. Fruit bats often feed in groups, sometimes flying long distances to reach fruiting trees.

The world's many bat species are usually grouped according to their eating habits. Besides fruit-eaters, there are bats that specialize in

↓ CHILDREN AT GHARIAL POOL

flower nectar, insects, small mammals, frogs, birds, fish, and, of course, blood. The bats that you're most likely to encounter in the United States are insectivorous and are very helpful in keeping summer mosquito populations under control.

As you leave JungleWorld, stop to take a look at the "jungle countdown" clocks. One of these two digital counters ticks off the number of acres of rain forest remaining (the number decreases by 100 acres per minute), and the other tallies up the world's human population, which increases at the rate of 180 per minute. It is a sobering thought that, with each passing minute, more and more people are crammed onto a shrinking planet. These stark facts remind us that, unless we act to control our numbers and protect wild places, wildlife will soon be just a memory.

> On board the Bengali Express, take a two-mile ride through the wilds of Asia and meet Siberian tigers, Asian elephants, and Indian rhinoceros.

WILD ASIA (E9–F5)

Covering over 38 acres in the zoo's southeastern corner, Wild Asia offers an introduction to the exotic wildlife of the Far East. Begin your journey at Wild Asia Plaza, a re-creation of a bustling Asian bazaar. In the summer, under the pagoda-roofed pavilions, you'll find refreshments, gift shops, and other amenities, including rest rooms, telephones, and water fountains. The open-air Dragon Theater is the site of educational programs and animal behavior demonstrations—check the posted schedule for show times. And before you leave Wild Asia Plaza, you may want to experience one of the earliest forms of human transportation by taking a camel ride. All the activities and services in Wild Asia Plaza are seasonal, open from May through October.

↑ MONORAIL

→ INDIAN RHINOS

As you walk through the plaza, you'll pass JungleWorld on your right. This large indoor exhibit is home to animals from the warmer regions of Asia, as well as smaller species, including tropical birds and insects. Ahead and to your left, you'll see the Bengali Express Monorail station, where you'll begin your wildlife-watching safari through the

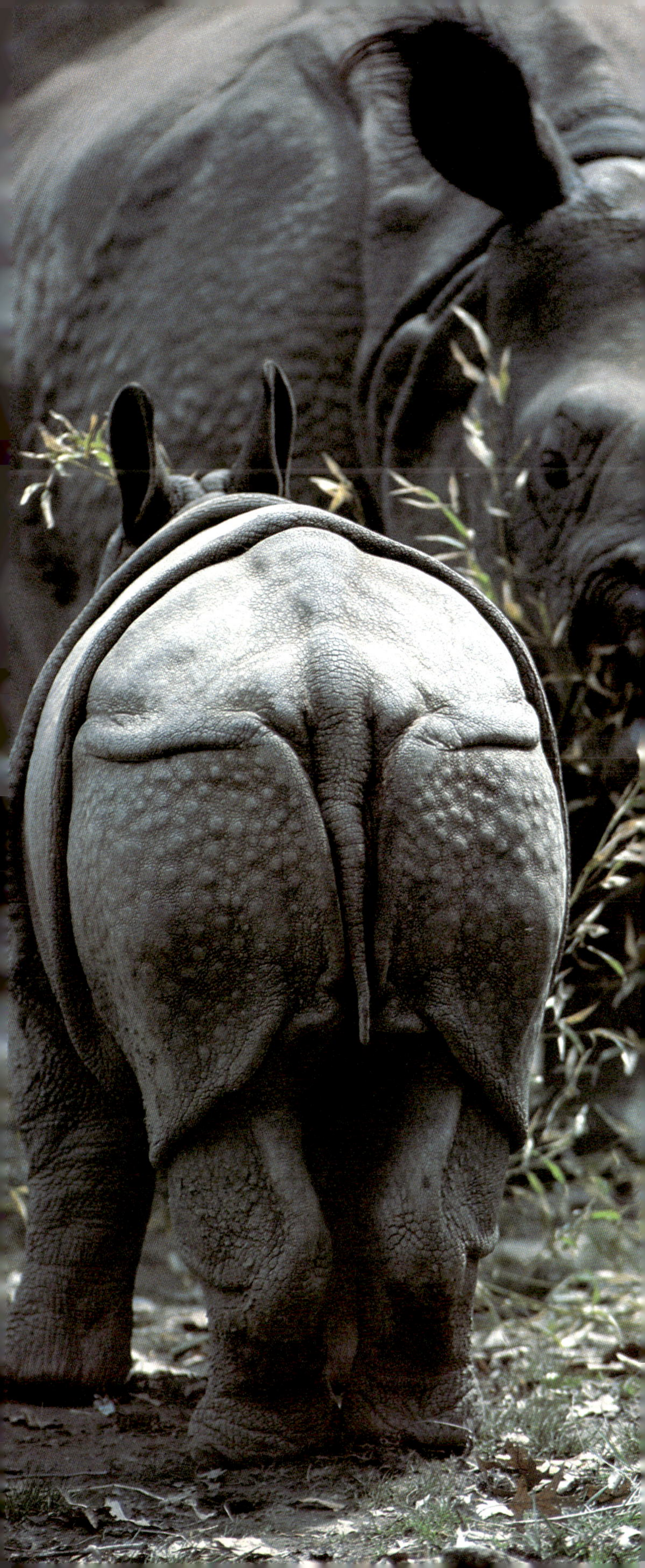

meadows, mountains, and forests of the Far East. Like the Wild Asia Plaza services, the Bengali Express operates seasonally, from May through October.

After buying a ticket, you'll board a specially designed monorail that's open only on one side, so that every rider has an unobstructed view of the animals. In the next twenty minutes, you'll pass through a variety of Asian habitats, each home to its own characteristic wildlife. Your tour guide is an experienced animal spotter—he or she will direct your attention to interesting animals and wildlife babies that you might otherwise miss. There will be plenty of opportunities for photos, and your tour guide will be happy to point out the best spots. The Bengali Express covers nearly two miles, at speeds between 2 and 6 miles per hour. As the scenery changes, so does the elevation: from ground level to thirty-five feet above the Bronx River.

After you cross the river (here renamed the Irrawaddy after a river in Burma), your train passes through central India's Kanha Meadow. Here you'll spot blackbuck, a species of antelope that inhabits the open plains. The blackbuck is one of the few antelopes in which males and females look radically different. The males are a rich, dark brown with striking white hair on the underparts and white circles around the eyes. Topping it off are impressive spiral horns that may reach 27 inches in length. The females, by contrast, are a yellowish fawn color and lack horns—altogether a less imposing sight than the males.

Sharing the Kanha Meadow with the blackbuck are barasingha deer and axis deer. Barasingha, also called swamp deer, are often found in marshy grassland and are highly endangered in the wild due to habitat loss. You'll also see axis deer, which can be identified by their white spotted coats. Axis deer are also rare in many parts of their natural range.

After leaving the meadow, your train will climb into the South China Hills, home to the beautiful and gravely endangered Formosan sika deer, a species that has been extinct in the wild since 1973. The Bronx Zoo herd is descended from sika deer that arrived at the zoo in 1940.

Next comes Cambodia's Angkor Forest, where you will see gaur, the largest of the world's wild cattle. Both male and female gaur are dark reddish-brown with conspicuous white "stockings" on all four legs. The

↑ GAUR

→ FORMOSAN SIKA DEER

PAGES 64 & 65:

SIBERIAN TIGER

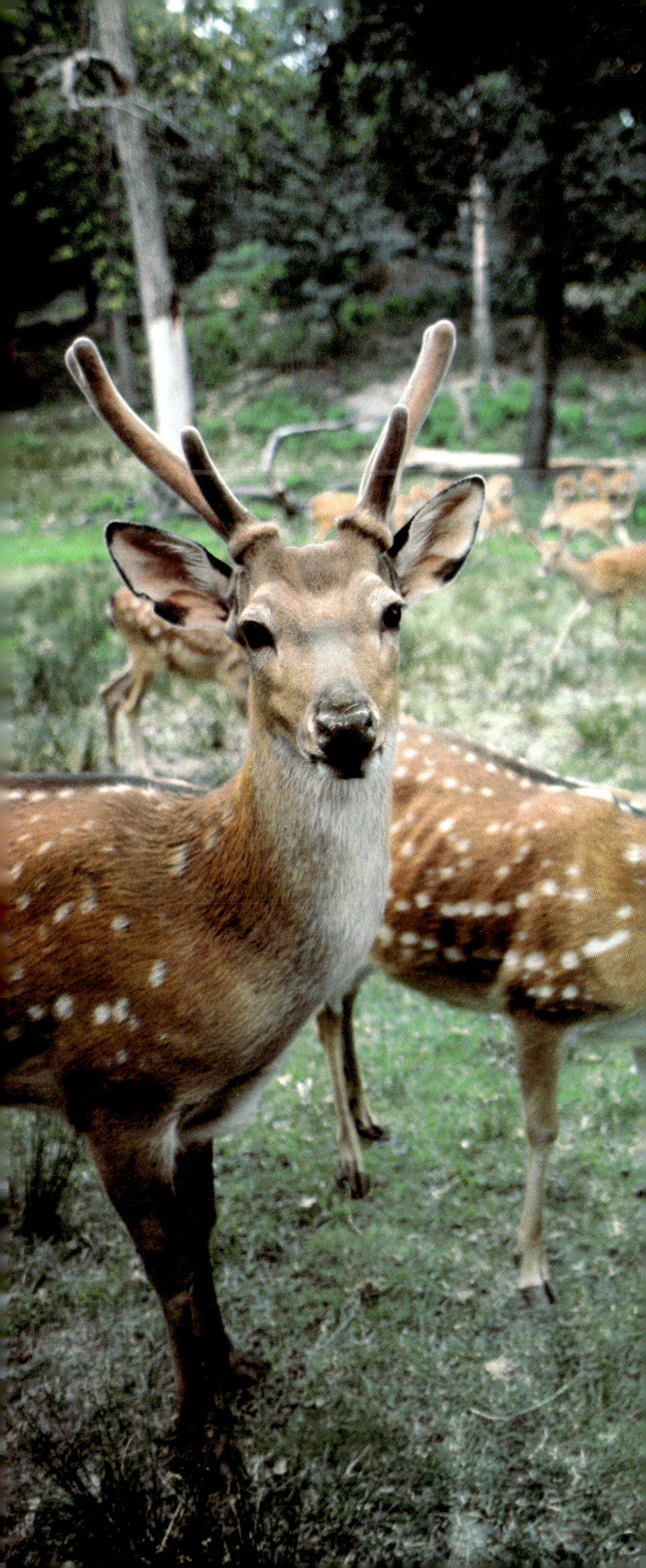

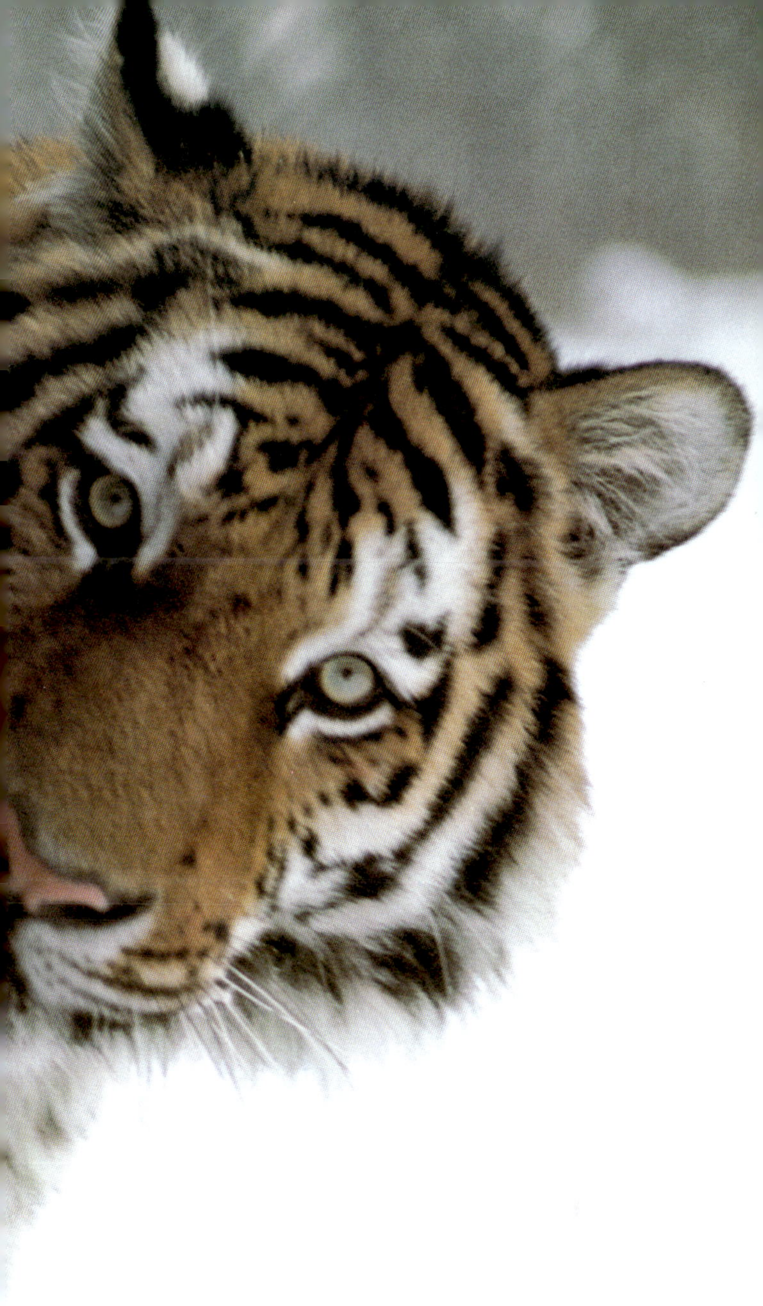

males can be distinguished by their larger size (they can stand six feet tall at the shoulder) and the prominent hump over the shoulders. Gaur are forest dwellers, but they seek out grassy clearings where they can graze on fresh grasses and leaves. Like many other inhabitants of the Asian forests, gaur are threatened by destruction of their habitat. And as people and livestock move into formerly wild areas, gaur become vulnerable to diseases transmitted by domestic cattle.

After leaving the forest, your monorail approaches the Tiger Machan, named for a type of treetop perch or blind from which tigers were once hunted. Here you'll see the largest of all the great cats: the Siberian tiger. All tiger subspecies, including the Siberian tiger, are seriously endangered in the wild. In fact, there are now more Siberian tigers breeding in zoos than in nature. They have lost both undisturbed forest land and their prey to human expansion, and they are also illegally hunted for their skins, bones, and other "parts," which are often used in traditional Asian folk medicines. Given these threats, it is not surprising that tigers are vanishing. Establishing protected areas is one way to protect these magnificent cats, but even a large tiger reserve can support only a small number of tigers. The Wildlife Conservation Society is at the forefront of efforts to save tigers in the wild.

↑ SIBERIAN TIGER

Leaving the tigers, you'll revisit the Kanha Meadow, where you may spot peafowl and babirusa, an extraordinary type of wild pig whose big tusks grow through its palate and out the top of its face!

In the next habitat, Thailand's Khao Yai Reserve, you may see Asian elephants bathing in their Olympic-size pool. This species is also threatened in its natural habitat. To learn about efforts to protect elephants, be sure to visit the Keith W. Johnson Zoo Center.

Next, you'll pass through Nepal's Chitawan Valley, a heavily wooded lowland exhibit where you'll view another of the largest land animals on earth, the Indian rhinoceros. Like their African relatives, In-

↑ BABIRUSAS

↑ ASIAN ELEPHANTS

dian rhinos have been hunted almost to the brink of extinction for their horns, which have long been thought to have medicinal powers. Until recently, whole rhino horns were also much in demand for use as handles of the traditional "jambia" daggers carried by men in North Yemen. As with elephants, one of the keys to protecting rhinos is to eliminate the demand for these products.

After leaving the Chitawan Valley, your train climbs into the hilly Rajasthan Uplands of northwestern India. Here you may glimpse a herd of sambar deer browsing on leaves and grass. Sambar are Southeast Asia's largest deer and the males carry stout antlers that can grow to a length of over three feet. In the wild, sambar deer are most active at dusk and dawn.

Before your train pulls into the station, you'll pass through the Karakoram Range, where you'll see tahr, a wild goat that comes from the mountainous Himalayan region, and the red panda, a relative of the familiar backyard raccoon.

> Bats in the World of Darkness navigate by listening to the echoes of their ultrasonic calls, the way submarines use sonar.

WORLD OF DARKNESS (D7)

For humans, who are active during the day and rely very heavily upon their eyesight, a visit to the World of Darkness can be a bit unsettling. Entering the building, you are plunged into darkness, and it's only after your eyes adjust to the dim lights that you begin to see some of the building's many nocturnal inhabitants. Unlike diurnal (day active) animals, most nocturnal animals lack the ability to see color and have eyes that are more specialized for night vision. They may also depend much more on their senses of touch, smell, hearing, or a combination of these.

As you walk through the building, you will be struck by the large number of bats on exhibit. There are nearly 1,000 species of bats around the world, and nearly all of them are active only at night. If you spend a few minutes

↑ SHORT-TAILED FRUIT BAT

watching the spear-nosed bats, flying foxes, and other bats in the exhibit, you'll be amazed at their agility in dodging trees and other obstacles.

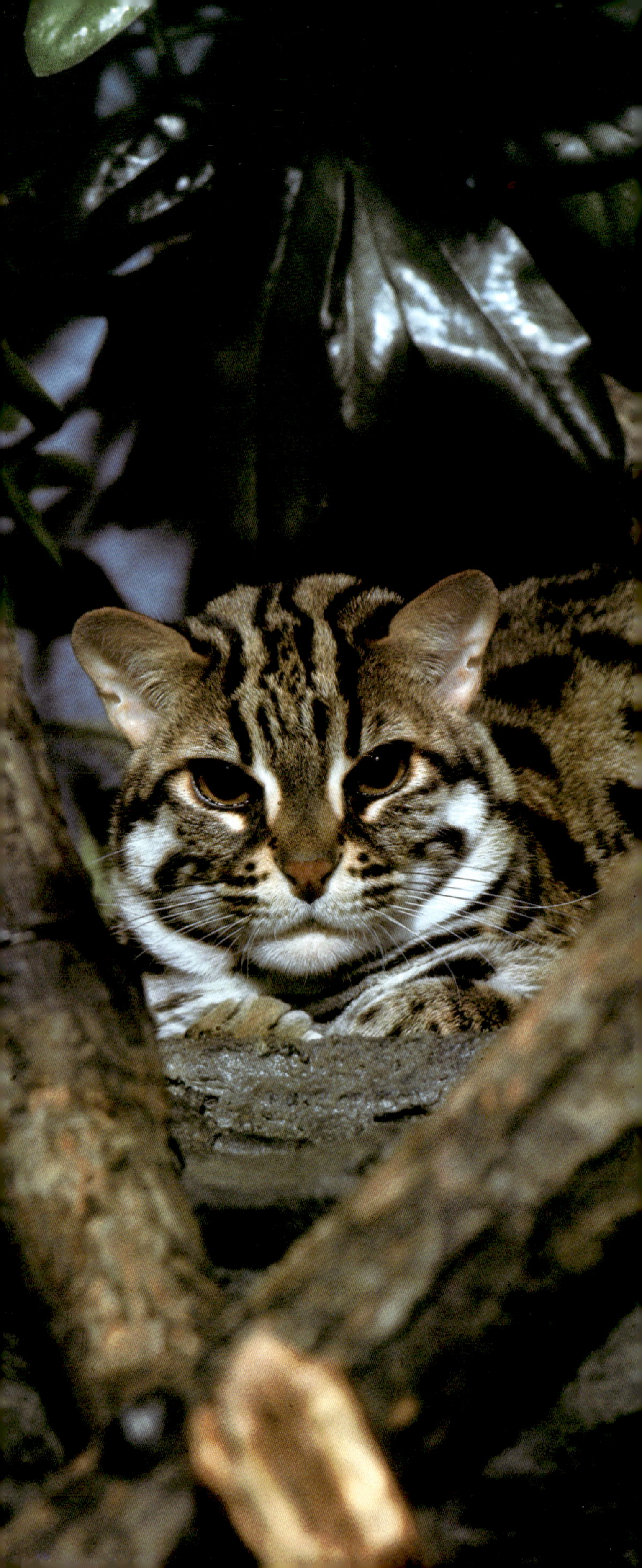

Most of these bats navigate at night and find food by producing a series of short ultrasonic calls that bounce off objects in the environment and are picked up by the bat's extremely sensitive ears. Some fruit-eating bats, like the endangered Rodrigues fruit bat, also have a keen sense of smell, which helps them home in on food.

Vision is of little importance to the naked mole-rats, small, nearly sightless rodents that spend their entire lives in underground burrows. Naked mole-rats are the only mammals known for an insect-like social organization. Within a mole-rat colony, which may number over 80 individuals, there is one breeding female (similar to the queen in a bee hive), a few breeding males, and a large number of nonbreeding workers whose job is to keep the burrows clean and care for the young. Mole-rats make their way through the tunnels mainly by touch—the few hairs that remain on their mostly naked bodies are extremely touch-sensitive. They also have well-developed senses of hearing and smell. In fact, researchers believe that chemical signals, perhaps transmitted as an odor in the breeding female's urine, play a role in suppressing breeding in the worker mole-rats.

↑ NAKED MOLE-RAT
← INDIAN LEOPARD CAT

One nocturnal animal that does rely on vision is the galago, or bush baby, and one of the most striking features of this little animal is its enormous round eyes. Although they use vision to get around, it is far less effective as a means of communication. To signal to one another, bush babies use sound (loud cries) and smell (urine marking).

↑ GALAGO

Like most other cats, the leopard cat spends its days asleep and its nights on the prowl. This small spotted cat hunts by climbing into a tree and leaping onto its prey, which includes birds, rodents, and other small mammals.

Although mammals make up most of the inhabitants of the World of Darkness, birds, reptiles, amphibians, and smaller creatures are also represented. Before you leave, try to spot a boa constrictor or a glow-in-the-dark scorpion. And when you get home tonight, be on the lookout for nocturnal animals in your own neighborhood. Skunks, opossums, and raccoons are just a few of the creatures that may be prowling around your yard at midnight.

BIG BEARS (D6)

Surrounding a high ridge of natural rock just up a path from the Himalayan Highlands are polar bears and grizzly bears. Despite their designation as land animals, polar bears are the most aquatic of all the bear species. In the wild, they spend many hours near water hunting for the ringed seals that are their primary food. If necessary, a polar bear can swim for hours through frigid water to get from one ice floe to another, propelling itself with its partially webbed feet. To protect against the cold, the polar bear has a thick fat layer and water-repellant fur coat. The white fur provides camouflage to a hunting bear and, in addition, helps to keep the bear warm. Although apparently white, the hairs are transparent, hollow tubes, which channel the sun's warmth to the bear's skin. Underneath the fur, the bear's skin is black, helping to retain heat. Further adaptations to the cold include fur-covered feet and small ears to minimize heat loss.

↑ POLAR BEAR
↙ GRIZZLY BEARS

The grizzly bear is a race of the widespread brown bear, which includes the great Kodiak bear—largest of all bears. Brown bears are found in Alaska, Canada, parts of the continental United States, western Europe, Russia, and northern Japan. Most grizzly bears are, in fact, dark brown, but fur color is quite variable and can range from light tan to black. In some individuals, the long hairs over the shoulders and back are tipped with white, giving the animal a "grizzled" appearance—hence the name grizzly bear.

The grizzly bear has been called the most dangerous animal in North America—next to humans. The grizzly's reputation has much to do with reproduction. A female grizzly bear will usually produce no more than 6 to 8 young during her lifetime, so the survival of each cub is vitally important. A mother grizzly defends her cub vigorously, taking even the smallest threat very seriously. Many people assume that male grizzlies fight to protect their young. In reality, however, the male's aggression serves a different purpose: to ward off other males who might mate with an available female.

> Six generations of snow leopards — one of the most elusive and beautiful of the world's big cats — have been bred at Himalayan Highlands.

HIMALAYAN HIGHLANDS (D6)

Designed to resemble a remote mountain forest habitat in Nepal, the Himalayan Highlands is home to the highly endangered snow leopard, as well as red pandas, Temminck's tragopan (a kind of pheasant), and white-naped cranes.

Shy, solitary, and often nocturnal, the snow leopard is one of the most elusive and least known of the world's big cats. It lives in remote, mountainous habitats, moving to different altitudes with its migrating prey, which includes wild sheep, ibex, musk deer, and smaller mammals such as hares, mice, and birds. The soft, thick fur that protects the snow leopard from the cold has made this cat the target of hunters. Although they are protected in many countries, snow leopards are still killed for their skins and because they are considered a pest by shepherds. The Bronx Zoo began a snow leopard breeding program in the 1960's. Since then, more snow leopard cubs have been born in the Bronx than at any other American zoo. The breeding program, along with field studies sponsored during the 1970's and 1980's, have established the Wildlife Conservation Society as one of the world's leaders in snow leopard conservation. Snow leopards born at the zoo have been sent to many zoos around the United States, as well as to zoos in Australia, Canada, England, Russia, and Japan.

Just down the hill from the snow leopards are the red pandas, which resemble furry red raccoons. Red pandas are, indeed, related to raccoons, though they were once thought to be related to the bear-like black-and-white giant pandas. Both share some common characteristics: they are herbivorous, feeding mainly on bamboo; they have an enlarged wrist bone that is used like an opposable thumb to help the animal grasp food; and they are most active at night. Clues to the red panda's nocturnal habits include its large ears and well-developed whiskers, as well as the fact that it spends most daylight hours asleep in the trees.

↑ RED PANDA

Another inhabitant of Asia's high-altitude forests is Temminck's tragopan, one of the world's most beautiful and brightly colored pheasants. As with most other pheasant species, the male is much more colorful than the female, whose somewhat drab plumage helps to camouflage her as she tends to her eggs and chicks. Tragopans are polygamous (one cock will mate with several hens), and the males do not help with nest-building or care of the young.

↑ TEMMINCK'S TRAGOPAN

→ SNOW LEOPARD

One of the most beautiful of all crane species, the white-naped crane is highly endangered in nature as a result of habitat destruction. In an attempt to save these magnificent birds, Wildlife Conservation Society scientists have pioneered methods for breeding cranes in captivity. One innovative approach utilized an artificial egg containing a thermometer, motion sensors, and a radio transmitter. When placed into the nest next to a real egg, the imposter egg recorded important data about the cranes' incubation behavior. This information was then used to design a new and more effective artificial incubator.

NORTHERN PONDS (C5)

Walking north from the Himalayan Highlands, you'll pass the Northern Ponds, home to a wide variety of waterfowl, including trumpeter swans and several species of ducks. Indigenous to North America, the trumpeter is the largest and the rarest of the swans. These imposing birds were abundant throughout the Northwest and Canada until the 1800's. But, like the bison, trumpeter swans fell victim to westward-moving settlers, who killed thousands for food and feathers, which were used to make powder puffs. By the 1930's, fewer than 100 trumpeter swans remained. Relocation of birds to other areas, coupled with the discovery of an Alaskan population of trumpeters, has saved this species from extinction. Trumpeters are now protected throughout their range and are slowly increasing.

↑ CHILD VIEWING WATERFOWL

↑ TRUMPETER SWAN

Trumpeter swans live in lakes, marshes, and flooded grasslands. They build large mound-like nests of moss and grasses and lay from three to seven eggs. Young swans, called cygnets, are grayish-brown in color, attaining their pure white adult plumage when they are about one and a half years old.

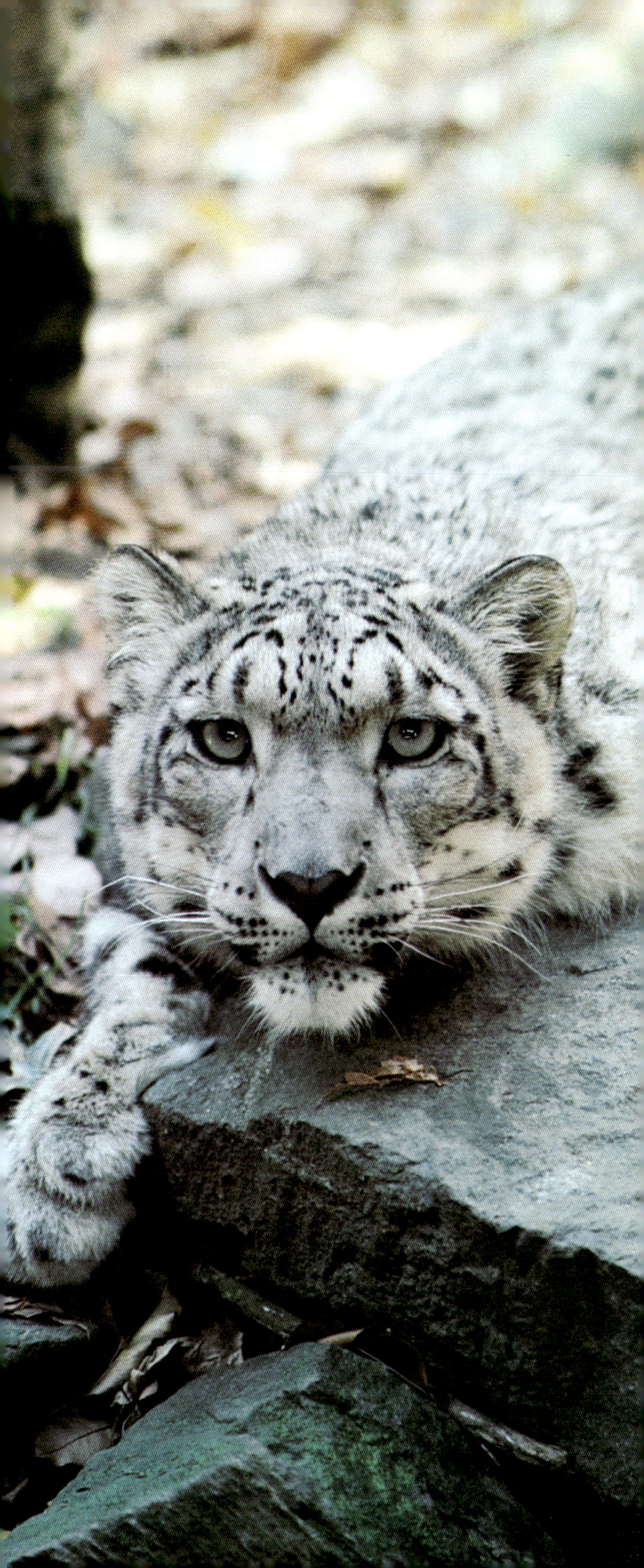

MEXICAN WOLVES (C4)

Next to humans, rats, and domestic animals, wolves have found their way to more places in the Northern Hemisphere than any other land creature. They can adapt to subfreezing conditions on the Arctic tundra, to the hot, dry deserts of the Middle East, and to almost every environment in between.

Wolves are pack animals and survive by banding together to hunt and kill prey that is often much larger than themselves. It's not uncommon for a pack to bring down an animal weighing ten times as much as an individual wolf. If cattle or other domestic animals are available, wolves may target them because they are easier to capture. This behavior has made wolves extremely unpopular with ranchers. In fact, Mexican gray wolves like those seen at the zoo are extinct in the southwestern United States as a result of persecution from cattle ranchers and government hunters. In an attempt to save the Mexican wolf, zoos around the country are cooperating with the governments of the United States and Mexico on a breeding and reintroduction program. The first captive-bred wolves were released into the wild on March 30, 1998.

Their adaptability and highly developed social behavior are two factors that made it possible to domesticate wolves and produce dogs, a process that began more than 12,000 years ago. Although today's dogs look very different from wolves, they still share behaviors with their wild ancestors. If you watch Mexican wolves, you're likely to recognize behaviors much like those you've seen in your own dog at home.

PERE DAVID DEER (C4)

Like the Mongolian wild horse, the Pere David deer is extinct in nature and now exists only in zoos. In fact, Pere David deer had already disappeared from the wild in 1863, when a captive herd was discovered behind the walls of the Imperial Hunting Park near Beijing by Father Armand David, a French Jesuit missionary. This herd was wiped out during the Boxer Rebellion in the early 1900's, leaving only eighteen animals that had been brought together two years earlier in England. In 1946, the Bronx Zoo acquired four deer from England, and those four were the founders of our current herd. Many of the deer born at the Bronx Zoo have been sent to other zoos to help establish new herds and thus perpetuate this ancient species.

↑ PERE DAVID DEER
→ MEXICAN WOLF

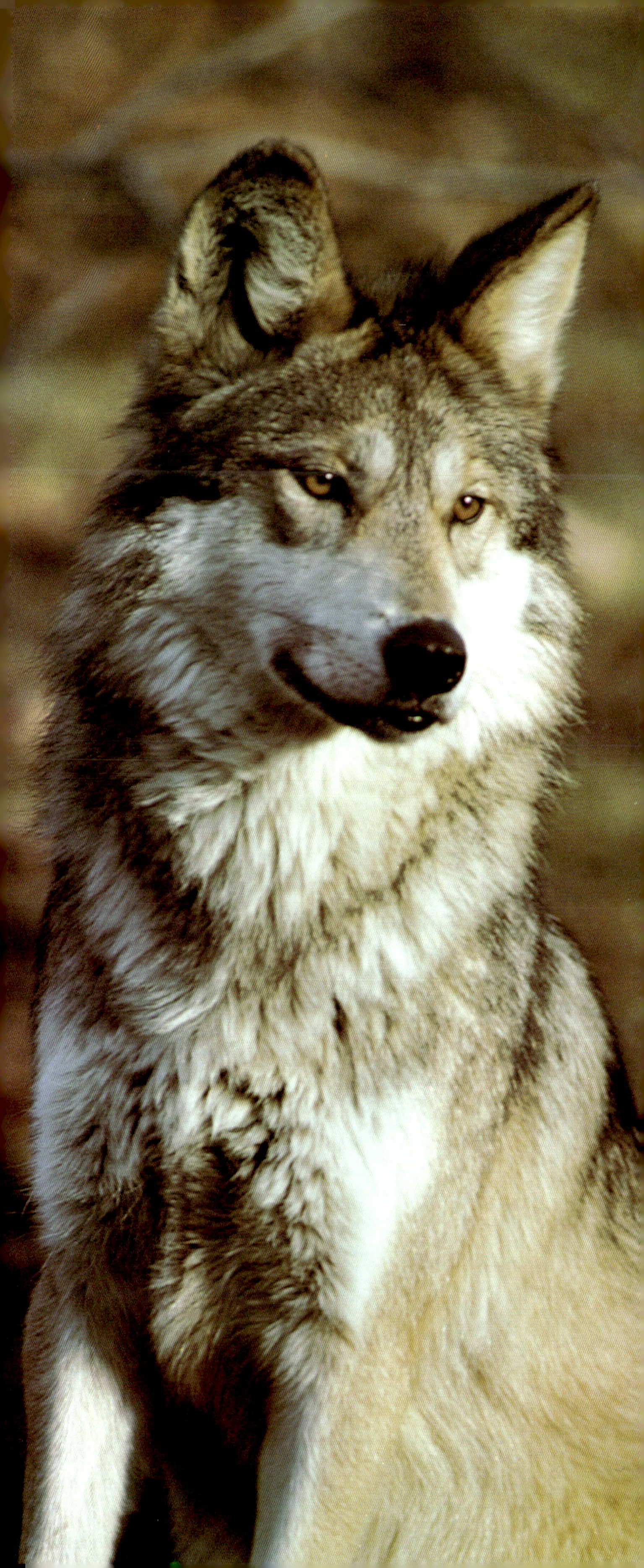

WORLD OF BIRDS (D3)

The World of Birds, which occupies a hillside near the zoo's Bronx Parkway entrance, displays many bird species in exhibits that closely resemble their natural habitats. A wide variety of ecosystems are represented, including a native New York woodland and African and South American rain forests.

Different bird species vary tremendously in size, shape, appearance, and behavior, but the first thing you'll learn when you enter the World of Birds is what all birds have in common: feathers. As you move through the building, you'll learn more about eggs, nests and nest-building, courtship and mating, and many other fascinating bird facts.

↑ COCK OF THE ROCK

Although the World of Birds features birds from all corners of the earth, the real focus is on birds of the tropics. The most spectacular exhibit in the building is the South American rain forest, where you'll walk through the treetops, and where an occasional thunderstorm drenches the forest and keeps the humidity high. As in the other naturalistic exhibits, patience is key if you want to see everything this exhibit has to offer. Take a few moments to stand quietly, watch, and listen, and you're likely to spot more birds than you ever imagined.

> At the World of Birds, many exhibits are barrier free, with no separation between visitors and spectacular toucans, bee-eaters, and sun bitterns.

Some of the most beautiful inhabitants of the World of Birds are birds of paradise, which are native to New Guinea. In addition to their brilliantly colored feathers, which often have a metallic or velvety sheen, the birds usually have long decorative plumes and tufts that can be raised at will. As in many other species of birds (and some mammals), only the males are so stunningly attired. The females are inconspicuous, with feathers of dull brown. In the wild, the usefulness of this fantastic plumage becomes apparent when a large number of males gather together in a tree and perform exotic "dances" while the females look on. Some birds of paradise hang upside down from a branch and dance while spreading out their iridescent feathers. The combination of the dance, the spectacular plumage, and the calls the birds sometimes utter during the dance makes a bird of paradise display something truly remarkable.

↑ BIRD OF PARADISE: NEST, CHICKS, ADULT

The bird of paradise's gorgeous plumage has, unfortunately, put these birds in jeopardy. During the 1800's, bird of paradise feathers were in much demand to adorn ladies' hats, and uncontrolled hunting resulted in decimation of many bird of paradise populations in New Guinea. It was only in 1924, after long efforts by the Wildlife Conservation Society, that the taking or possession of these birds was declared illegal. They are now protected throughout their range.

Other fascinating residents of the World of Birds are hornbills, large birds that at first glance can be mistaken for toucans. (Despite their large beaks, toucans are actually more closely related to woodpeckers than they are to hornbills.) The hornbill's beak is often brightly colored and is surprisingly light in weight for its size, because it is filled with a sponge-like network of horny material. Residents of the tropics, hornbills are found in trees, where they feed on ripe fruit, insects, and small vertebrates. Desirable fruit is often found at the ends of slender twigs, making it all but inaccessible to these large, heavy birds. In such instances, the long, lightweight beak extends the bird's reach, enabling it to grasp fruit that would otherwise be out of bounds.

↑ GREAT HORNBILL

↑ HELMETED CURRASOW

Hornbills are also known for their unusual breeding behavior. After selecting a suitable nesting hole, the female goes inside and proceeds to seal up the entryway with wood pulp, food, and other materials. She leaves just a tiny slit through which the male delivers food for her and the chicks. The female may remain inside the nest for several months, first incubating the eggs and then caring for the young until they are old enough to go out on their own.

Another inhabitant of the tropical forests, the helmeted currasow, can be recognized by the bony outgrowth on its forehead. Currasows have short, rounded wings and usually glide from tree to tree, flapping their wings only when necessary. These birds are also known for their loud and piercing calls. Some currasow species have an unusually long trachea, which increases their vocal power.

Among the most interesting tropical birds you'll see in the World of Birds are the parrots and cockatoos, including the spectacular palm cockatoo. This large bird is a native of Australia and New Guinea and, like many other parrots and cockatoos, is becoming increasingly rare in its natural habitat. The Wildlife Conservation Society is working to preserve endangered parrot and cockatoo species through conservation efforts in the field and captive breeding programs here at the zoo. Behind the scenes at the World of Birds are propagation facilities for cranes, pheasants, birds of paradise, and many other endangered avian species.

↑ PALM COCKATOO
→ AMERICAN BISON

BISON RANGE (D3)

Until their virtual extermination in the late 1800's, an estimated 60 million bison roamed the central plains of North America. By 1890, the population had plummeted to only a few hundred animals. But thanks to the efforts of the founders of the Wildlife Conservation Society, then known as the New York Zoological Society, the bison was saved, and more than 100,000 of these imposing animals now live on protected refuges in the American West.

Bison herds are constantly on the move in search of grass, which is their chief food source. For most of the year, the herds consist of females and young, with males joining the groups only during the late summer breeding season. There is intense competition among the males for receptive females, and fierce fights, punctuated by violent head-ramming and loud roaring, are frequent occurrences during this time. After about ten months, the calves are born. Each calf remains with its mother for close to a year. Some cows are ready to breed again the summer after giving birth, while others won't breed until the following year.

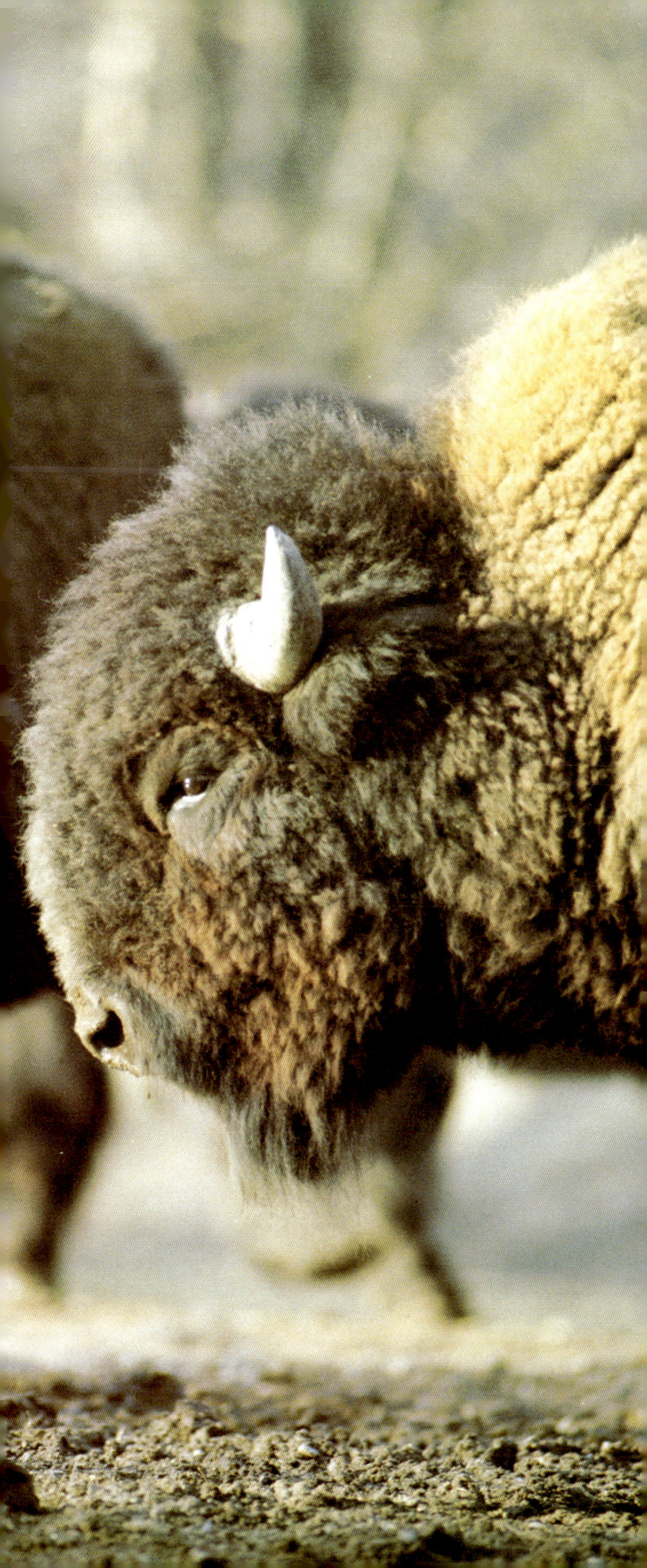

Part Two
BEHIND THE SCENES

KEEPING THE ANIMALS HEALTHY

Caring for the more than 7,000 animals that make their home at the Bronx Zoo is a demanding job and one that requires the input of a wide variety of animal care professionals. Curators, wild animal keepers, veterinarians, nutritionists, and scores of other specialists work together to insure the health and well-being of the zoo's mammals, birds, reptiles, amphibians, and invertebrates. In major areas (mammalogy, ornithology, and herpetology), one or more curators preside over a large team of animal keepers, collections managers, supervisors, and maintainers. These individuals are responsible for the day-to-day care of the animals, including feeding, cleaning and maintaining exhibits, moving animals when necessary for breeding or other reasons, and keeping a sharp eye out for signs of illness or other potential problems. The animal care staff is frequently involved in enrichment programs designed to offer extra stimulation to the zoo's wildlife.

↑ RED PANDA CUB
→ CURATOR WITH BABY GORILLA

In addition to overseeing the daily care of the animals, the curators cooperate with other accredited zoos on breeding programs (more about this under Bronx Zoo Breeding Programs). The curators also work with other zoo staff on the design of new zoo exhibits and the redesign of outdated exhibits, and they often conduct field research in their particular areas of interest.

Maintaining a healthy Bronx Zoo animal collection is the responsibility of all involved in animal care, but most specifically, it is the job of the veterinarians at the Wildlife Health Center. The new Center, on the grounds of the Bronx Zoo, has been the Wildlife Conservation Society's central health care and health research facility since 1985. Departments of clinical medicine, pathology, and nutrition care for the animals at the Bronx Zoo, as well as those at the Society's other New York City wildlife parks and the Wildlife Survival Center in Georgia.

Responsibility for the daily veterinary care of thousands of fish, amphibians, reptiles, birds, and mammals falls to the Department of Clinical Studies. Here, veterinarians provide advanced medical and surgical attention, including endoscopy, laser surgery, microsurgery, dentistry, ultrasonography, radiology, and intensive care. An in-house clinical laboratory, pharmacy, surgical scrub room and sterile prep room,

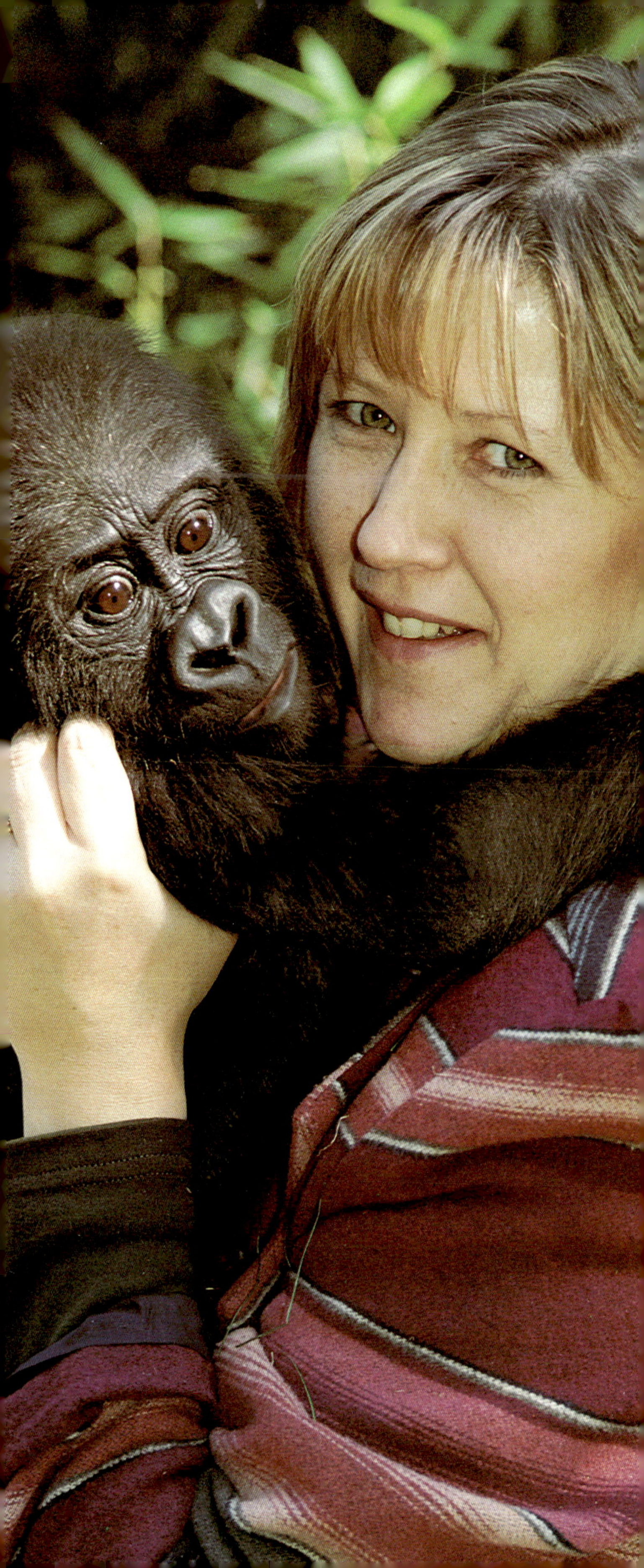

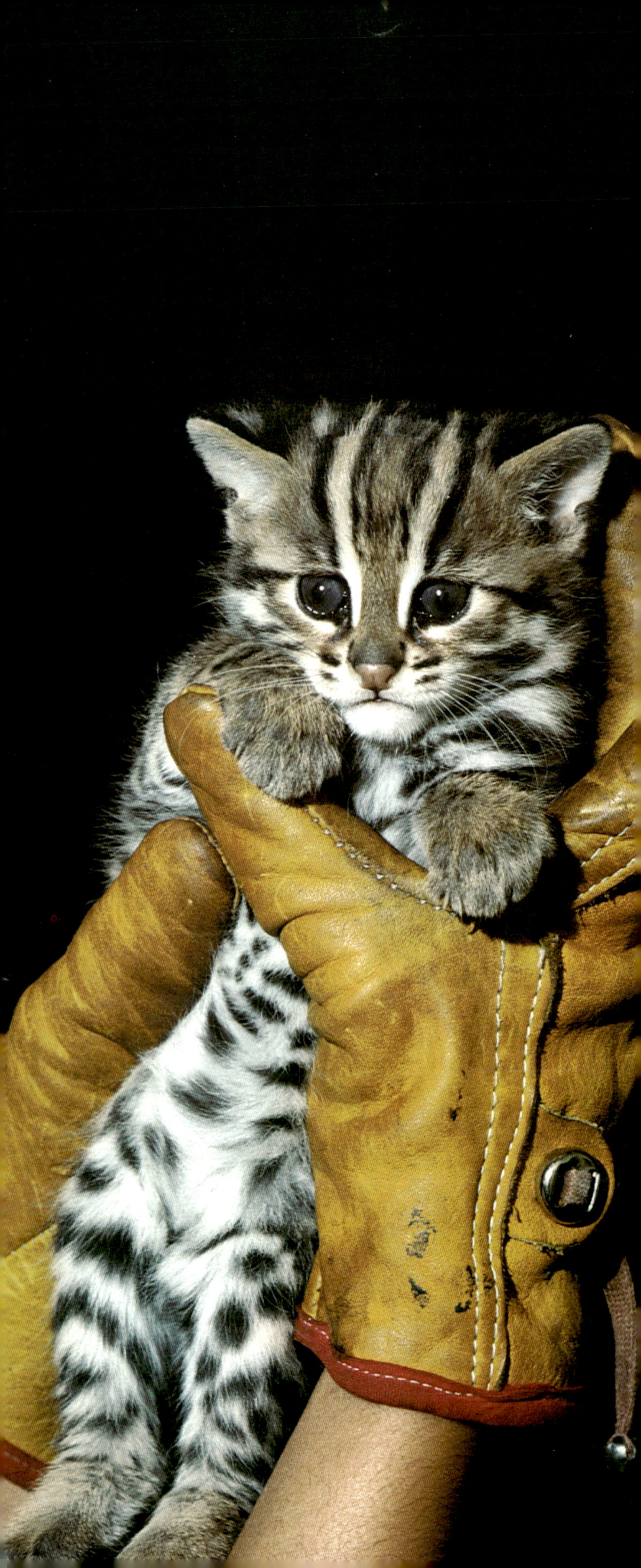

imaging suite, surgical suite, recovery area, and clinical wards support the busy clinical practice.

The Department of Pathology provides the foundation for the work of the other Wildlife Health Center departments. Here, tissue samples from animals at the zoo and wildlife centers, as well as from wild animal populations, are stored for study, helping veterinarians to identify – and ultimately treat – a wide range of diseases affecting wildlife.

To combat wildlife disease and health problems in the field, WCS created the Field Veterinary Program. While WCS conservationists pursue the goals of conserving wildlife populations for the future, many wild animals struggle with pressures imposed on them today. The Field Veterinary Program augments ongoing conservation efforts by dealing with current health problems as well as providing direct medical help to WCS's scientists in the field.

↑ CHINESE ALLIGATOR HATCHING
↑ EXAMINING SNOW LEOPARD
← LEOPARD CAT KITTEN

Back in New York, the Department of Nutrition, created in 1986, oversees the diets of animals in the five city wildlife facilities and at the Wildlife Survival Center in Georgia. Diet preparation, nutritional assessment, feedstuff handling and storage protocols, and quality control are all coordinated by Nutrition Department staff.

Maintaining the nutrition of zoo animals is a challenge, because it is rarely possible to provide the same foods that these animals would obtain in the wild. It is often possible, however, to duplicate the nutrients contained in the food. In the Bronx Zoo-based nutrition lab, one of the few such facilities in the world, hundreds of food items are analyzed for nutrient content. Many of the foods sent in for analysis have been collected in the wild by WCS field biologists.

MAKING IT ALL WORK

Caring for the zoo's animal inhabitants is a big job, but it is only one of the many things that must done to keep the Bronx Zoo up and running. Other "behind the scenes" activities include updating existing exhibits and designing new ones, caring for the wide variety of plants that grow both outdoors and in the exhibits, running the zoo's many restaurants, snack bars, and gift shops, and performing the construction and maintenance work needed to keep the 265-acre park looking its best.

The Exhibition and Graphic Arts Department (EGAD for short) plays a vital role in the planning, design, and fabrication of zoo exhibits and brings together experts from WCS's diverse departments with outside experts. Whether the job is to construct a hillside habitat for baboons or to fabricate miles of vines for mandrills and marmosets, EGAD's first task is to create stimulating naturalistic environments for animals. Equally important, however, is the creation of compelling places for zoo visitors. In every exhibit, EGAD attempts to immerse the visitor in the animal's world in order to inspire people to care and take action to conserve wildlife.

> More than 400 employees work 365 days a year to keep the Bronx Zoo one of the premier wild animal parks in the world.

Although wildlife is the star at the Bronx Zoo, the lush plantings, both indoors and out, play an important supporting role. In each exhibit, the Horticulture Department has made every effort to simulate the animals' natural environment, including plants that would normally be found in their native surroundings. In many cases, this means growing and maintaining tender tropical plants that would normally not survive in New York City. Even when they can be grown successfully in the zoo's nursery or greenhouses, exhibit plantings need frequent replacement. Each year approximately 15,000 plants and almost 5,000 tons of grass seed are added throughout the park, and over 1,500 plant species can be seen throughout the zoo.

↑ EXHIBIT DESIGNERS WITH MODEL

→ CONGO CONSTRUCTION SITE

A large group of Maintenance and Construction tradespeople, almost 100 strong, work 365 days a year to keep the Bronx Zoo's grounds and buildings in shape to accommodate more than 2 million visitors annually. The Guest Services team handles food, merchandise, parking and admissions, and ride operations throughout the park. On certain summer days, they will take care of the needs of as many as 40,000 visitors!

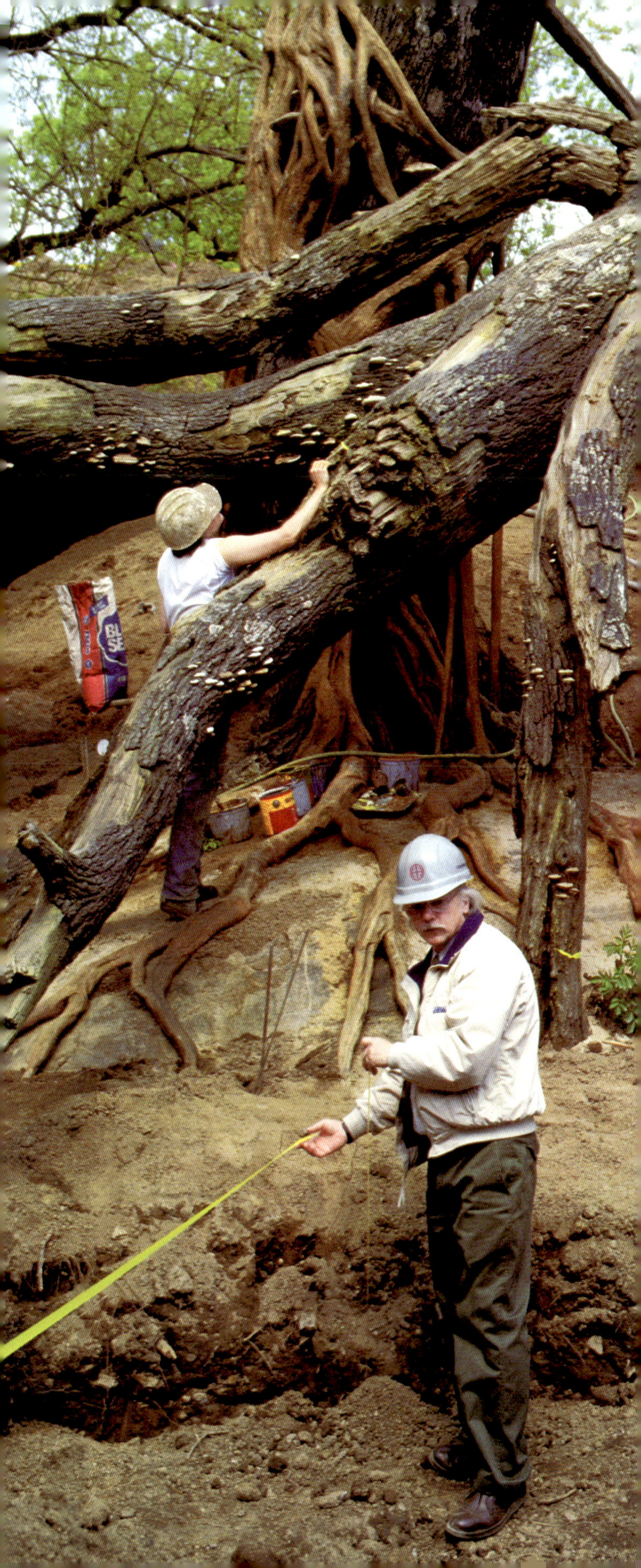

EDUCATION

Introducing people to the world of wildlife is the goal of the Bronx Zoo's Education Department, and whether you're eight or eighty, you're bound to find a wildlife adventure to your liking. School science curricula developed by the Bronx Zoo's Education Department are used by thousands of teachers and their students nationwide and in several foreign countries, including China and Papua New Guinea. These curricula, including Pablo Python Looks at Animals (grades K-3) and Wildlife Inquiry through Zoo Education (WIZE) (grades 7-12), make science fun by integrating classroom lessons and zoo-based field trips.

For organized school groups, the Education Department offers highly participatory programs that include activities in the classroom and at the zoo's world-famous exhibits, as well as a chance to touch armadillos, boa constrictors, and other exotic creatures. The department also offers a wide range of professional development opportunities for teachers of grades K-12. Teacher workshops help educators to bring new excitement to their classrooms by showing them how a focus on animals can enliven the subjects they are teaching.

↑ GIRL WITH ARMADILLO
↑ EDUCATION CLASS

Kids as young as three can have a close encounter with rabbits, owls, and prairie dogs in a program at the renowned Children's Zoo. Older children can learn about wildlife careers, such as zoo veterinarian, animal behaviorist, conservation biologist, or zookeeper. Families can sleep at the zoo overnight in a tented camp near the African market or join the Education Department's expert instructors for exciting programs that teach them about lions, gorillas, and other Bronx Zoo stars. There are courses just for adults, too, from an in-depth tour of the Congo Gorilla Forest to a romantic Valentine's Day dinner for couples in JungleWorld's lush rain forest. For more information about Bronx Zoo Education programs, please call (718) 220-5131.

↑ EDUCATION CLASS
→ ZOO CAMP

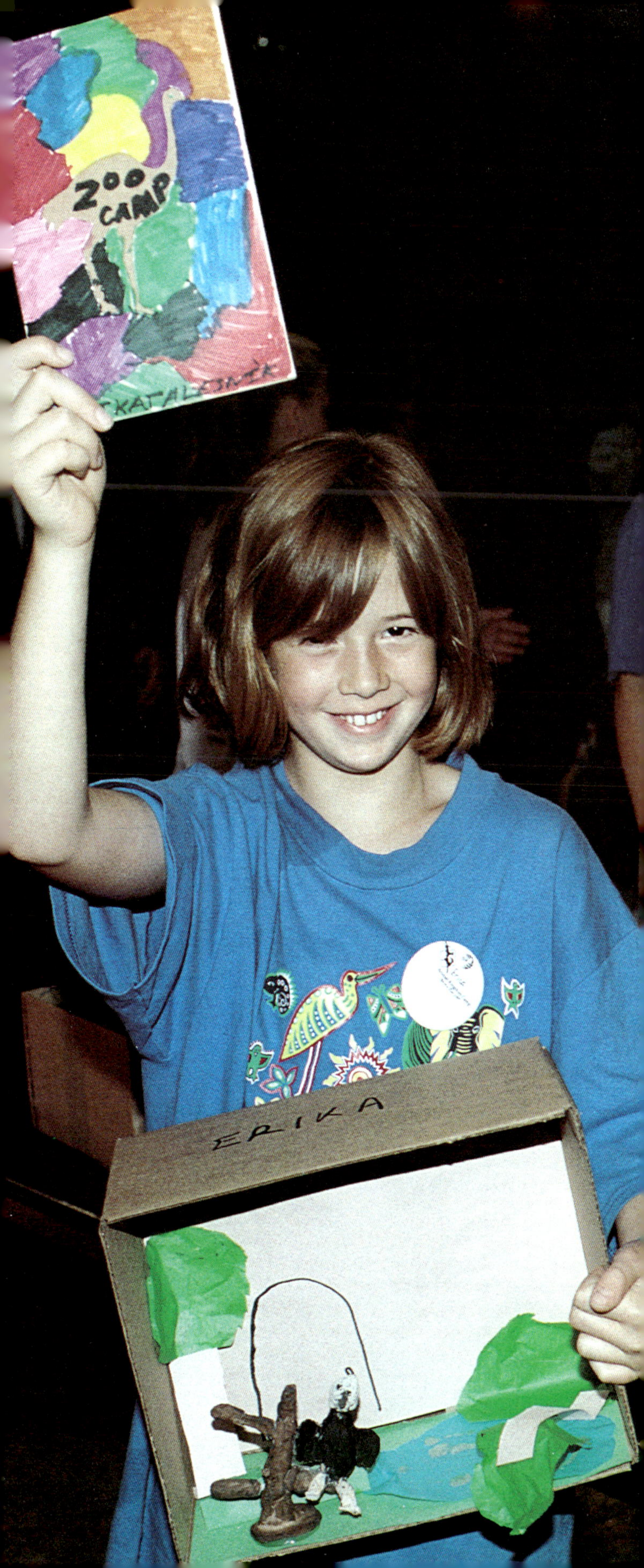
ZOO CAMP
ERIKA

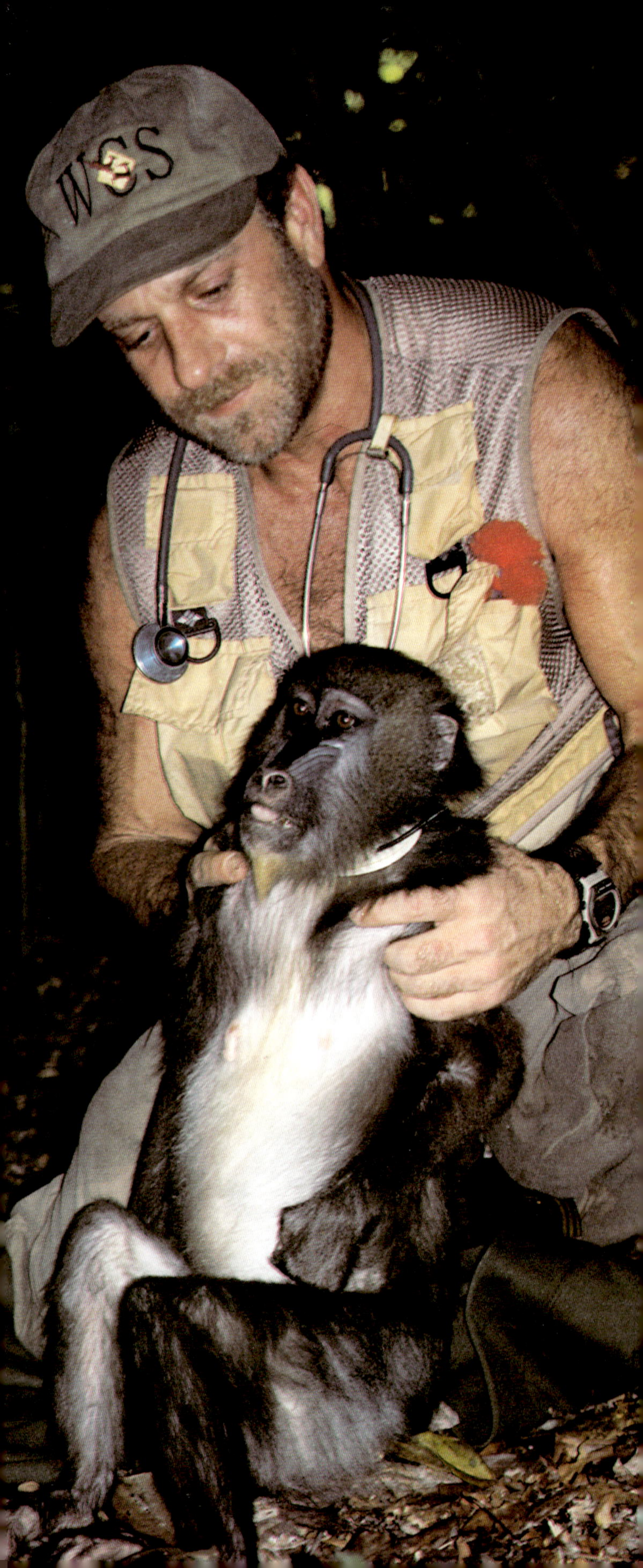
WCS

SCIENCE RESOURCE CENTER

Headquartered at the Bronx Zoo, the Science Resource Center helps Wildlife Conservation Society biologists use the latest scientific information and techniques in their work saving wildlife. SRC is a learning center where scientists combine wildlife information, new technologies, and key wildlife sciences, such as behavioral ecology and conservation genetics, to come up with answers to specific research questions. Knowledge gained from such projects is used to find practical ways to save endangered species and to protect wildlife habitat.

An example of how WCS scientists use cutting-edge technology to learn about wildlife is a project that utilizes satellite tracking to study manatees. Since 1998, scientists have been tracking two West Indian manatees living in Belize, Central America. These large, plant-eating marine mammals are listed as endangered, and to help ensure manatee survival, scientific studies aimed at understanding their biology are needed. In this study, radio transmitters were attached to the animals, and several times a day the manatees' locations are broadcast to satellites. Each day, scientists process the data sent to them and prepare animal tracking maps. The results are expected to increase our understanding of manatee behavior and help wildlife managers in Belize to protect and conserve these magnificent animals.

The SRC is also home to the WCS Conservation Genetics program, which was started in 1989 in order to apply current techniques in molecular biology to the conservation of wildlife. This program uses genetic marking techniques to identify unique evolutionary lineages, distinct populations, and individual animals. In one project, DNA markers were used to construct a pedigree for black rhinos in Tanzania's Ngorongoro Crater. This information is extremely valuable to wildlife managers as a guide to moving rhinos among small, isolated populations. By taking a scientific approach to managing these highly endangered animals, it may be possible to maximize their breeding success and insure their future in the wild.

↑ ELECTRO-FISHING
← FIELD VET WITH MANDRILL

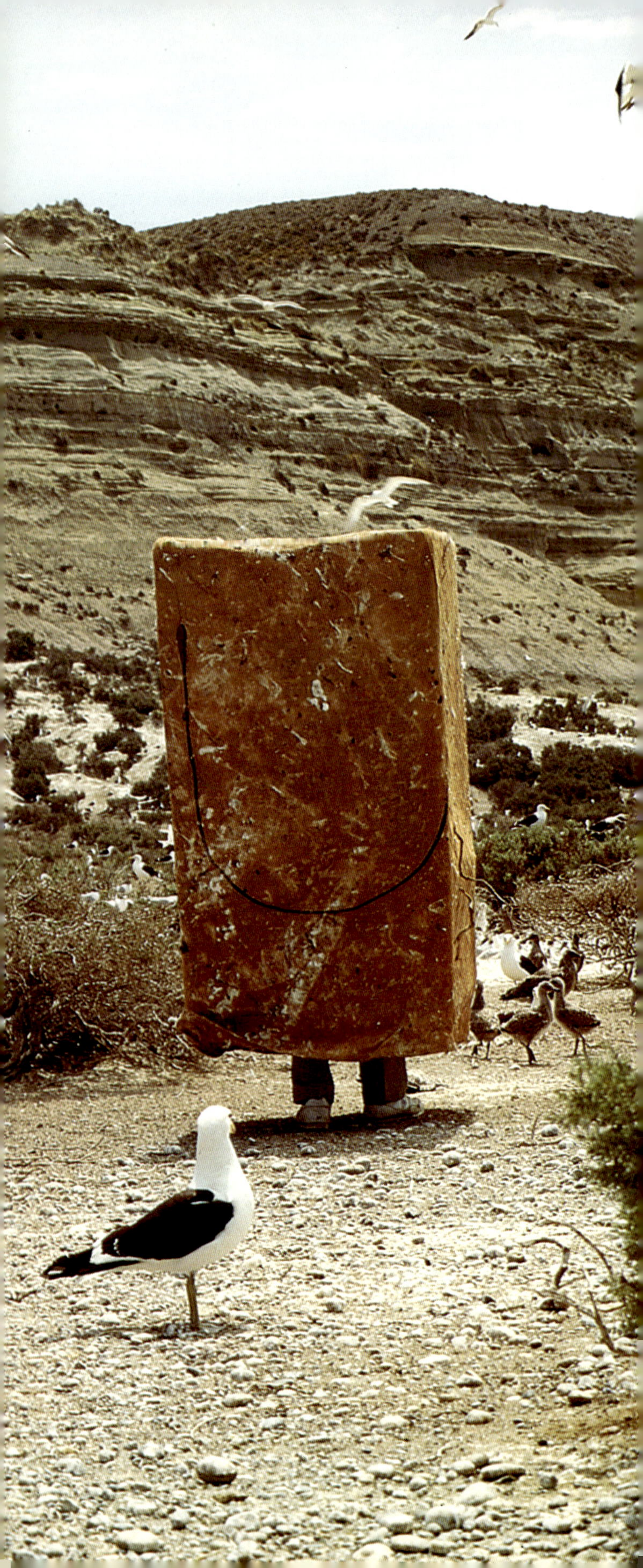

INTERNATIONAL CONSERVATION

In an attempt to gain protection for wildlife and wild places, the Wildlife Conservation Society is active around the world, working to preserve both wild animals and their habitats. With 65 staff scientists and over 100 research fellows, WCS has the largest professional field staff of any U.S.-based international conservation organization. The Society currently conducts more than 300 field projects in over 50 countries in Africa, Asia, Latin America, and North America.

Using a variety of research methods, WCS field staff gather information on a particular region to determine its need for protection. Wildlife surveys, in-depth ecological studies, computerized satellite geographic information systems, radio tracking, and human usage studies are just some of the ways WCS scientists determine the best ways to safeguard an area and its wildlife. After the data have been compiled and analyzed, WCS works with local governments to establish and manage protected areas and parks. It is this full-service approach — from collecting the basic data to working with the highest levels of government — that sets the WCS International Conservation program apart from all others.

↑ FIELD SCIENTIST IN INDIA
← WALKING BLIND

> Over the last decade, the Wildlife Conservation Society has helped save more than 100 million acres of wildlife habitat around the world.

The Wildlife Conservation Society has been involved in efforts to preserve wildlife in Asia since 1909, when Director for Tropical Research William Beebe conducted a pioneering survey of Asian pheasants. Today, burgeoning human populations and expanding economies threaten Asia's rich natural resources, and many animal species are heading toward extinction. In an attempt to save Asia's rich and exotic wildlife, WCS scientists have studied pandas in China, tigers in Thailand, and monkeys in Malaysia, to name just a few.

In Africa, the Society's protection efforts date back to 1920, when the Bronx Zoo's first director, William Hornaday, recognized the need to save South Africa's white rhinos. Forty years later, WCS Director for Science Dr. George Schaller pioneered studies of mountain gorillas in

Congo. Conserving Africa's wildlife presents a real challenge: unparalleled populations of exotic animals are quickly losing ground in a land faced with extreme poverty and the highest human population growth rates in the world. One of the most effective ways to protect Africa's wild animals is to set aside space for them, and, over the years, WCS has helped to establish many parks and protected areas throughout Africa. These include Tanzania's Ruaha and Tarangire National Parks, Kenya's Amboseli National Park, Congo's Okapi Wildlife Reserve, and, most recently, Madagascar's Masoala National Park.

Latin America is another area where human population growth is squeezing out wildlife – in fact, the human population of Latin America has tripled since 1950. Natural resource exploitation, including mining, agriculture, hunting, and fishing, puts further stress on already overused ecosystems. As a result, many animal species have been pushed to the brink of extinction. With over 100 projects in 17 countries, WCS is an important force for conservation in Latin America. These projects range from a cooperative effort with Peru's Machiguenga Indians to conserve endangered macaws to creating "Paseo Pantera" (Path of the Panther), an unbroken corridor of parks and refuges stretching the length of Central America. More than 90 percent of WCS's Latin American projects are run by nationals, who are in a unique position to understand local conditions and conservation opportunities.

Last, but certainly not least, WCS has renewed its century-old commitment to conserving the wildlife and spectacular habitats of North America. Our North American program is currently active in the East and the West, from studies of wolf reintroduction in the Adirondacks to the ecological role of large predators in the northern Rockies.

BRONX ZOO BREEDING PROGRAMS

The Wildlife Conservation Society sponsors hundreds of projects around the world designed to preserve wild animals in their natural habitats. But for many endangered and threatened species, it is not enough to save wild populations. Some, such as the Mongolian wild horse, are completely extinct in nature, with populations existing only in zoos around the world. For these and other animals, captive breeding programs are vital to the ongoing survival of the species.

↑ SIBERIAN TIGERS

→ GORILLAS

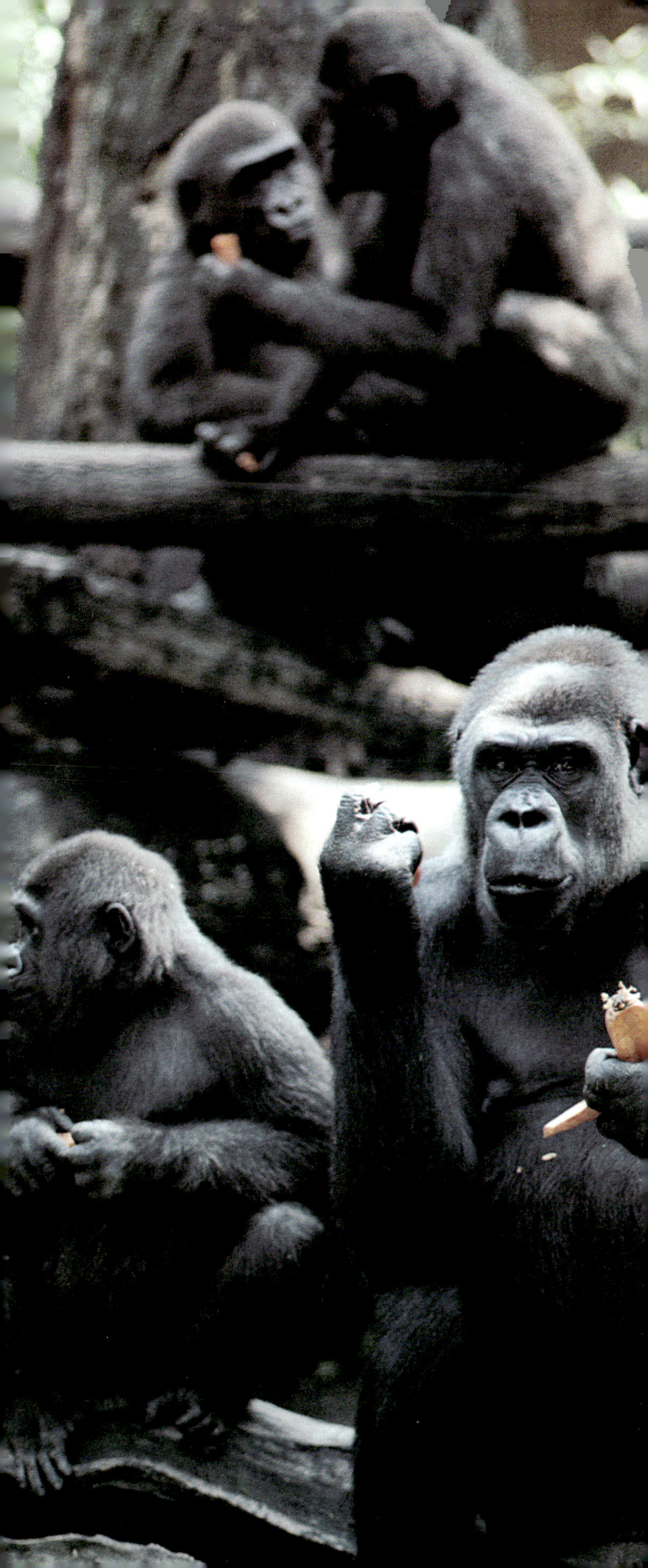

In order to coordinate breeding programs in zoos around the country, the American Zoo and Aquarium Association (AZA) has created Species Survival Plans for a number of the world's most highly endangered animals. The Species Survival Plan (SSP) is a cooperative effort among the AZA's 184 accredited zoos and aquariums and is designed to coordinate and carefully manage breeding efforts. The SSP is not a substitute for preserving animals in nature but one of many strategies aimed at sustaining healthy populations in captivity. These captive-bred animals could someday be candidates for return to the wild – if there are ever safe wild places to which they could be returned. By pooling their resources and cooperating, accredited zoos are making the most of their captive populations and insuring the future of species ranging from black-footed ferrets to palm cockatoos.

↑ SNOW LEOPARDS

> More than 2,000 animals are bred each year at the Bronx Zoo and the Wildlife Conservation Society's four other New York City parks, helping to maximize genetic diversity of important species.

WCS President and General Director William Conway was the father of the Species Survival Plan, and WCS staff members continue to be involved in carrying out SSP programs. One of the most important jobs is that of SSP coordinator for a particular species. The SSP coordinator is instrumental in deciding which individuals in a captive population should participate in breeding programs in order to maximize the genetic diversity of the species. WCS staff members serve as SSP coordinators for a number of species, including the Chinese alligator, snow leopard, and gorilla. Currently, WCS participates in over half of the established SSPs. In addition to SSP coordinator, staff involvement includes serving on management committees and acting as studbook keepers, who maintain historical records for a given species.

Among WCS's most successful breeding programs are those for snow leopards and gorillas. Dr. George Schaller, WCS Director for Science, was one of the first to study the elusive snow leopard in the wild, in the early 1970's. The Bronx Zoo's snow leopard breeding program began in 1966, and since then more than 70 cubs have been born in the Bronx, more than at any other American zoo. Dr. Dan Wharton, Director of the Central Park Wildlife Center, is the SSP coordinator for snow leopards. In this capacity, he keeps track of snow leopard populations in zoos around the country and makes specific breeding recommendations to each of the more than 50 participating zoos.

The Wildlife Conservation Society has also been a leader in gorilla conservation, both through field studies and captive breeding programs. The Bronx Zoo is home to one of the largest and most important breeding groups of western lowland gorillas in the United States. Our gorilla breeding program began in 1972, and since then more than 40 baby gorillas have been raised successfully. Dr. Wharton is also the SSP coordinator for gorillas and oversees the nationwide breeding program for this endangered species. You can see the zoo's breeding group of gorillas in the new Congo Gorilla Forest exhibit, which provides an unusually naturalistic and spacious environment for these primates, as well as many other animals from the forests of central Africa.

HISTORY OF THE BRONX ZOO AND WILDLIFE CONSERVATION SOCIETY

Since its founding in 1895, the Wildlife Conservation Society has had as its goals saving wildlife and inspiring people to care about and protect our precious natural heritage. In the 100-plus years that have passed, WCS has worked at both local and global levels to achieve these goals. Locally, WCS operates the largest system of urban zoos in the country. Our award-winning environmental education programs reach adults and children throughout the United States and in several foreign countries. And through our international conservation program, which operates over 300 research, training, and conservation programs on four continents, WCS is acknowledged as a world leader in saving endangered species and ecosystems.

↑ BRONX ZOO CIRCA 1910

It all began in the 1890's, when a group of prominent New Yorkers, including Theodore Roosevelt, lawyer Madison Grant, and paleontologist Henry Fairfield Osborn, decided that New York City needed a zoological garden. On April 26, 1895, the New York Zoological Society was chartered "to establish and maintain in said city a zoological garden for the purpose of encouraging and advancing the study of zoology, origi-

↑ SHIPPING BISON FROM ZOO TO WICHITA CIRCA 1907

nal researches in the same and kindred subjects, and of furnishing instruction and recreation for the people."

Just two years later, the society began its support of zoological research by sponsoring its first field project, a study of moose, caribou, mountain sheep and goats, bears, wolves, and other wildlife of Alaska and British Columbia. This study was followed by others that ultimately led to passage of the Alaskan Game Act of 1902.

In 1898, final plans for the Bronx Zoo were approved and building began in earnest. After a year and a half of intensive construction, the zoo opened on November 8, 1899, with 22 exhibit buildings and facilities completed. Construction has continued throughout the zoo's long history with improvements in the exhibits echoing those in zoological understanding.

The zoo's first director was William T. Hornaday, who had a deep interest in bison. It was this connection that got the fledgling Society involved in efforts to save the American bison in the early 1900's. Until the mid-1800's, as many as 60 million bison roamed the plains of the American West. But as settlers moved westward, these animals fell prey to hunters, who slaughtered them for their hides, for meat, or just for sport. By the end of the century, only a few hundred of these magnificent animals remained. Refusing to allow this symbol of the West to vanish forever, the New York Zoological Society, now the Wildlife Conservation Society, spearheaded an effort to save the bison. In 1907, the Society helped to establish a refuge, the Wichita Mountain Forest Reserve in Oklahoma, where bison could live free from hunters. Then the Society's founders shipped 15 bison from the Bronx Zoo out to Oklahoma, to help

> In the early 1900s, bison bred at the Bronx Zoo were sent west to rebuild herds on the verge of extinction.

found a new population. Thanks to the Society's efforts, the bison made an amazing recovery, and today, more than 100,000 of these animals live on America's western plains.

In the years that followed, the Society expanded its reach locally, nationally, and internationally, changing its name in 1993 to the Wildlife Conservation Society to reflect the breadth of its activities. Locally, the Society took over management of the New York Aquarium in 1902. The aquarium was then located in Manhattan's Battery Park, and it was not until 1957 that the present Aquarium for Wildlife Conservation opened on Coney Island. In 1916, the world's first zoo animal hospital was opened at the Bronx Zoo; it was later replaced with the modern Wildlife Health Center. The country's first formal zoo education program was organized at the Bronx Zoo in 1929. Today it reaches over 1.7 million school children in the New York metropolitan area and millions more across the United States and in several foreign countries.

Beginning in the 1980's, the Wildlife Conservation Society renovated and assumed operation of New York City zoos located in Manhattan, Queens, and Brooklyn. The first to open was the Central Park Wildlife Center. Following a complete redesign and renovation, the new facility opened in 1988, featuring animals from the three climatic zones: the tropics, the temperate regions, and the polar regions. In 1997, the Central Park Wildlife Center added a major new area: the Tisch Children's Zoo. The Queens Wildlife Center opened in 1992, as a contemporary showcase for spectacular American species. And in 1993, the Prospect Park Wildlife Center opened, as an all-new, interactive zoo that encourages children, especially, to actively learn about the world of wildlife. The combined yearly attendance at all the Bronx Zoo, Aquarium, and Wildlife Centers exceeds 4.4 million visitors. At these New York parks, and at the Wildlife Survival Center at St. Catherine Island, Geogia, WCS staff care for more than 10,000 animals representing 1,000 species.

On a national level, the zoo's founders quickly became involved in legislative efforts to protect animals. Bronx Zoo Director William Hornaday led a fight in the U.S. Congress to pass a bill banning the killing of migratory birds nationwide. Hornaday's cause was helped by the

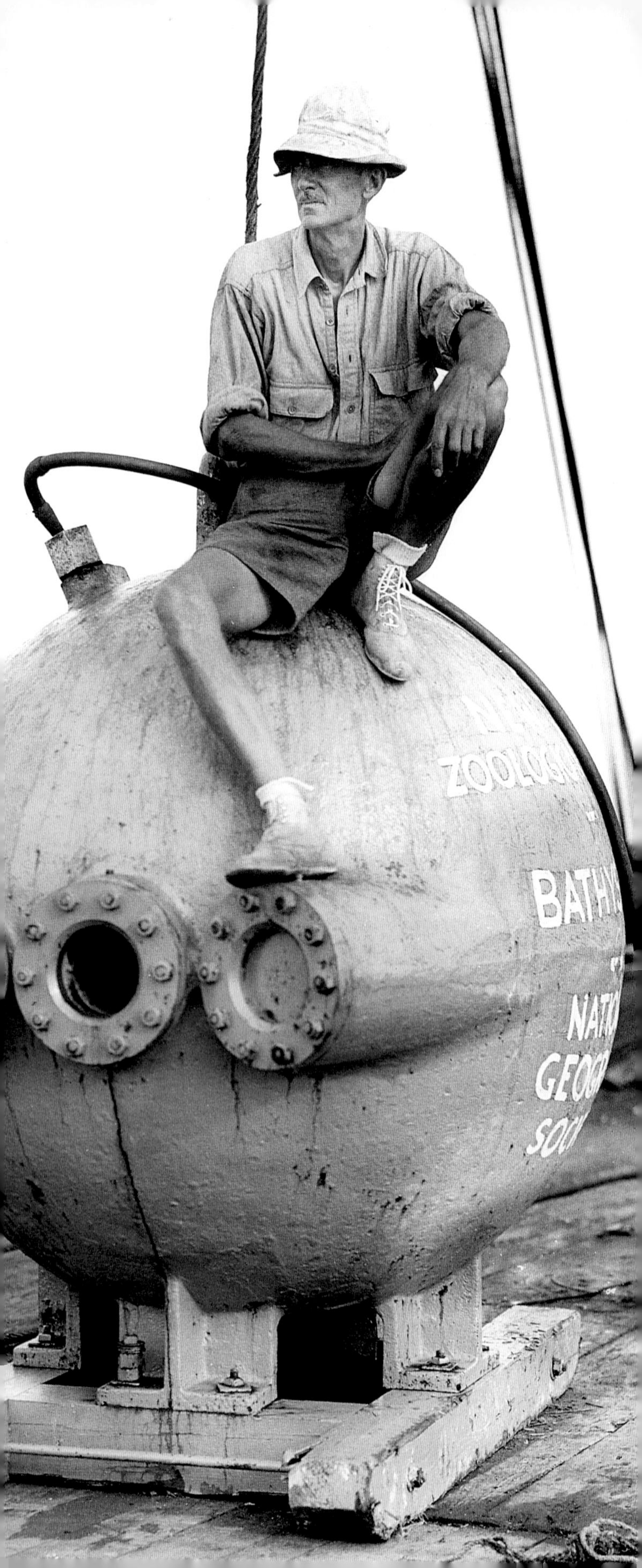

wide distribution of his book, *Our Vanishing Wild Life*, and in 1913, the McLean-Weeks Migratory Bird Law was passed. This was followed by the passage in 1929 of the Migratory Bird Conservation Act, also championed by the now-retired Hornaday. This federal law protected migrating birds by establishing fourteen bird sanctuaries around the country. Over the years, WCS has continued to be active in working for legislation to protect animals. As part of the six-member Ocean Awareness Campaign, WCS is currently working to change fishing regulations to protect sharks, bluefin tuna, swordfish, and other ocean fishes. And WCS efforts, especially meetings between WCS senior staff and federal government officials, were instrumental in preserving the Endangered Species Act.

The Society's field conservation program, which got its start with early expeditions to Alaska, continued to expand, eventually earning WCS a reputation as an international leader in field conservation. In 1909, Society scientist William Beebe traveled 50,000 miles throughout eastern Asia, conducting a comprehensive study of pheasants and related birds. This year-and-a-half-long odyssey culminated in the publication of his definitive work, *A Monograph of the Pheasants*. Some years later, Beebe made history by descending to a depth of 3,028 feet off the shores of Bermuda in a diving sphere only four feet in diameter. On this descent, he recorded a number of formerly unknown deep-sea species. Among the other WCS scientists who made history are Dr. George Schaller, who has conducted pioneering studies of giant pandas, Serengeti lions, tigers, and mountain gorillas; Dr. Alan Rabinowitz, whose remarkable studies of jaguars led to the establishment of a jaguar preserve in Belize; and Dr. Ian Douglas-Hamilton, who was instrumental in expanding our understanding of African elephants. These scientific studies have taught the world much about the behavior and ecology of these and other threatened species and have paved the way for conservation of these animals and their increasingly fragile environments. The Society's international conservation division now conducts more than 300 research, conservation, and training projects in over 50 countries around the globe.

As it has for the past 100 years, the Wildlife Conservation Society will continue to work at every level—local, regional, national, and international—to make people aware of the importance of wildlife and wild places, to save species and their environments, and to promote the cause of conservation.

← WILLIAM BEEBE ON TOP OF BATHYSPHERE

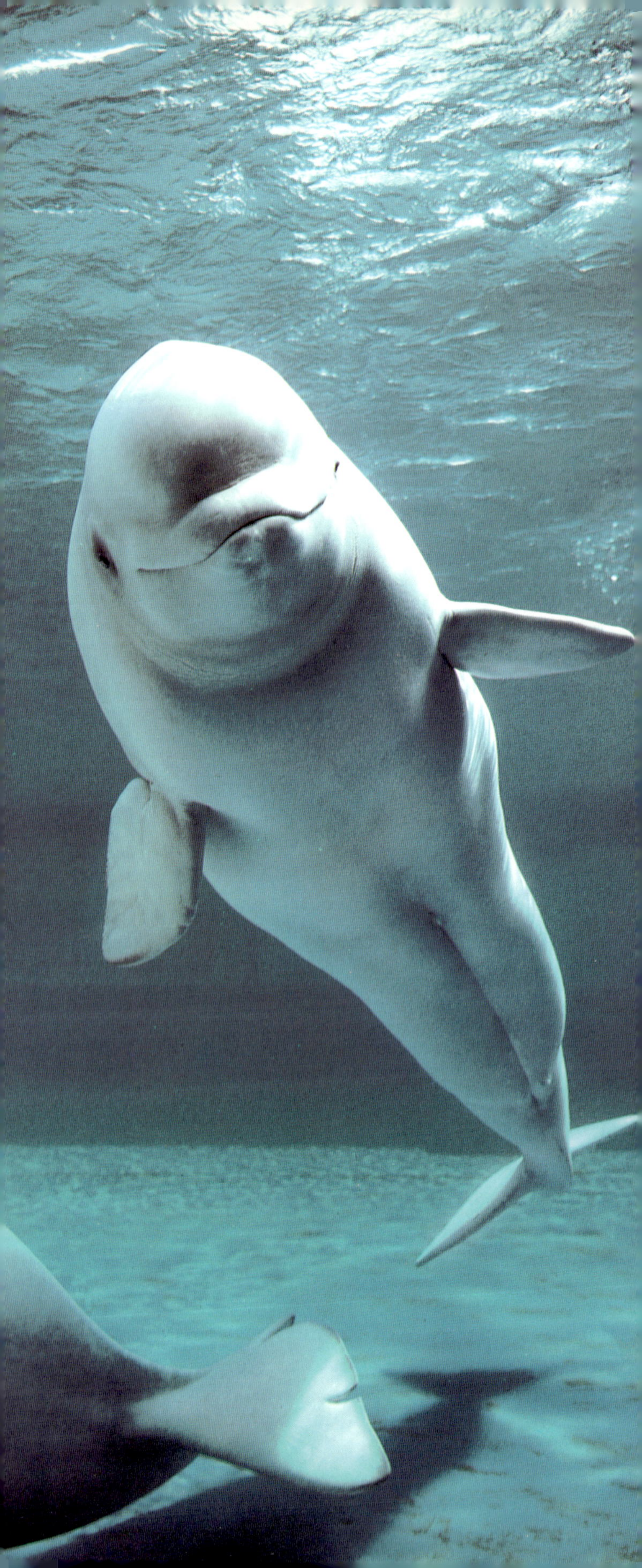

WCS PARKS THROUGHOUT NEW YORK CITY

NEW YORK AQUARIUM

Situated on 14 acres by the sea in Brooklyn's Coney Island, the New York Aquarium is home to thousands of fish and a multitude of marine creatures, including beluga whales, sharks, walruses, and dolphins. The aquarium is the longest operating facility of its kind in the country and continues to be a leader in aquatic science and conservation-related research. On your visit to the aquarium, be sure to visit Discovery Cove, which features hands-on exhibits and exciting video displays; Sea Cliffs, a 300-foot re-creation of a rocky Pacific coastal habitat; and the Aquatheater, where Atlantic bottlenose dolphins will astonish you with their intelligence and agility (May–September).

↑ WALRUS
← BELUGA WHALE

> Almost 4.5 million people visit the Wildlife Conservation Society's five wildlife parks in New York City.

CENTRAL PARK WILDLIFE CENTER

Tucked into only 5.5 acres in Manhattan's Central Park, you'll find the Central Park Wildlife Center. The main zoo is divided into three climatic zones: the Tropic Zone, an indoor rain forest environment that is home to free-flying birds and bats, endangered primates, red-bellied piranhas, and more; the Temperate Territory, where you'll see Japanese snow monkeys, red pandas, and California sea lions; and the Polar Circle, home to polar bears, harbor seals, Arctic fox, and an indoor Edge of the Icepack exhibit with gentoo and chinstrap penguins. The center also features the new Tisch Children's Zoo, where youngsters can explore a child-sized world of wildlife and meet friendly, familiar species.

↑ SEA LION
↑ POLAR BEAR

PROSPECT PARK WILDLIFE CENTER

Introducing children to the wonders of wildlife is the goal at Brooklyn's Prospect Park Wildlife Center, which includes both indoor and outdoor exhibit areas. In the World of Animals, youngsters can meet prairie dogs nose-to-nose, leapfrog across giant lily pads, walk among wallabies, and much more. In the Animal Lifestyles building, children can investigate communication among hamadryas baboons and the social life of cotton-top tamarins. The Animals in our Lives exhibits encourage children to observe and draw wildlife, including a busy family of meerkats. Outside, visitors can interact with friendly, touchable species in an inviting barnyard.

↑ HAMADRYAS BABOON
→ MOUNTAIN LION
↓ ANGORA SHEEP

QUEENS WILDLIFE CENTER

Pay a visit to the Queens Wildlife Center in Flushing Meadows Park for an American wildlife adventure. Here you will see spectacled bears, come face-to-face with a mountain lion, stroll through the open aviary, and discover majestic Roosevelt elk, American bison, and more.

Reopened in 1992 after extensive renovations, the Queens Wildlife Center showcases native American wildlife and features an interactive children's zoo, a barn and garden center where visitors can learn about animal and plant domestication, and an innovative education program.

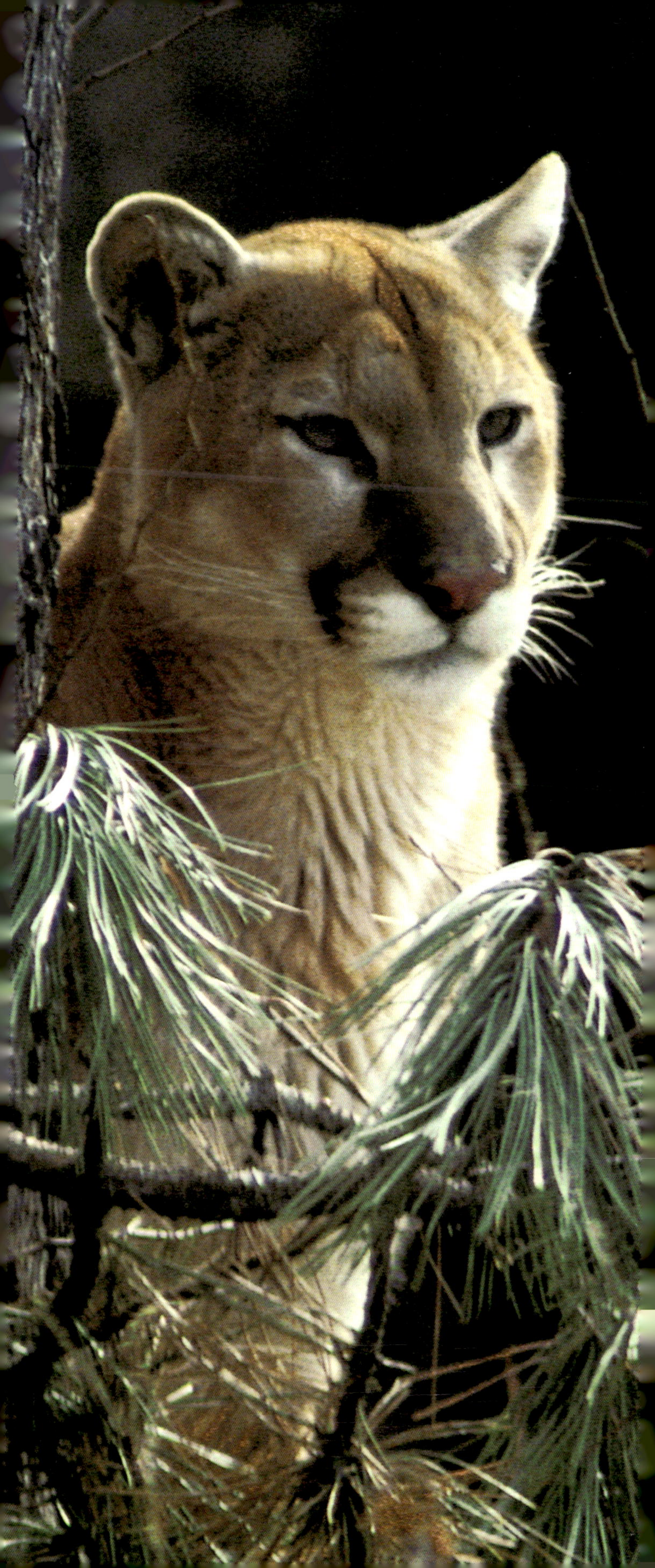

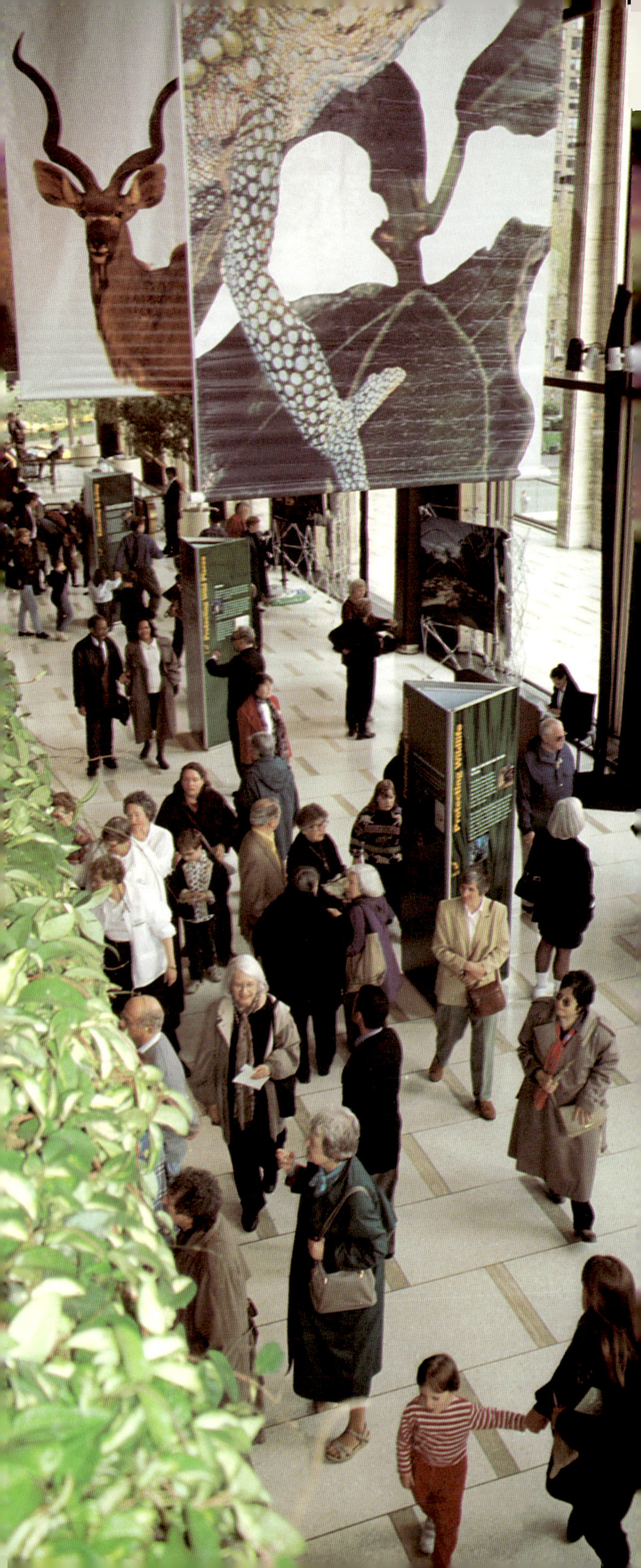
Protecting Wildlife

MEMBERSHIP, MAGAZINE, WEB SITE INFORMATION

Become a member of the Wildlife Conservation Society! As a member, you'll get free admission for one year to the Bronx Zoo; New York Aquarium; and the Central Park, Queens, and Prospect Park Wildlife Centers. You'll also receive:

→ a members' newsletter with information about special members-only events and activities at the wildlife centers.

→ one year of *Wildlife Conservation* magazine, a bimonthly publication filled with breathtaking photographs and articles that will keep you up-to-date on our worldwide conservation initiatives.

→ opportunities to travel with WCS scientists on exciting trips such as a family safari to Africa or South America.

→ invitations to lectures and special events and discounts on zoo education classes.

WCS membership categories include levels from Individual through Patron. Please call the WCS Membership office at (718) 220-5111 for more information.

Wildlife Conservation magazine seeks to inspire care for our precious natural heritage. This award-winning magazine, a benefit of membership in WCS, is also available on newsstands or by subscription by calling (718) 220-5121.

Visit us online at www.wcs.org! The Wildlife Conservation Society web page provides information on current WCS activities both in New York and around the globe. The home page changes weekly to reflect news items of interest, including scientific projects, special zoo and aquarium events, and updates from the world of conservation.

↑ WILDLIFE CONSERVATION MAGAZINE

← MEMBERS MEETING

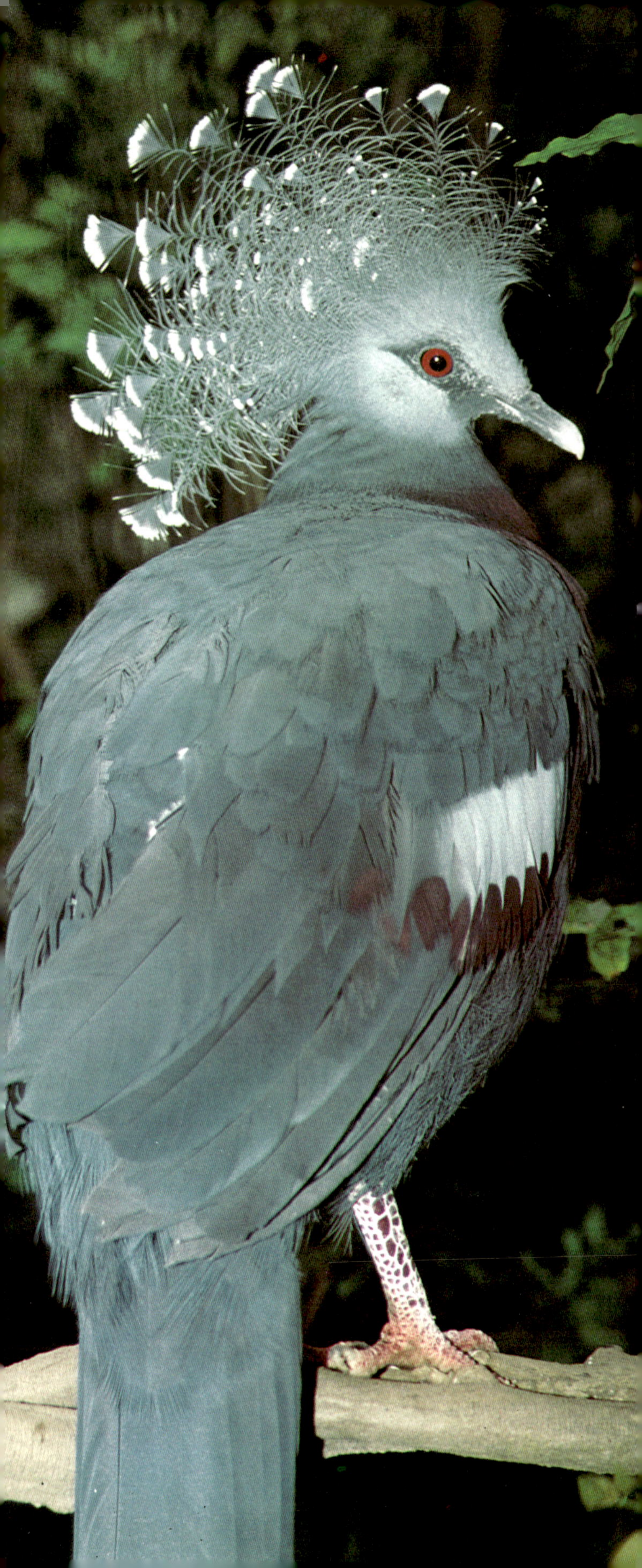

BRONX ZOO AT-A-GLANCE

Location: Bronx, New York

Year Founded: 1899

Size: 265 acres

Annual Attendance: more than 2 million

Employees: approximately 400

BRONX ZOO ANIMAL CENSUS (1998)

Mammals: 1,822 individuals of 130 species/107 births

Birds: 1,341 individuals of 272 species/145 hatchings

Reptiles/Amphibians: 1,291 individuals of 122 species/ 40 births & hatchings

Invertebrates: 2,025 individuals of 88 species

Children's Zoo: 604 individuals of 112 species/16 births & hatchings

Total Bronx Zoo: 7,083 individuals of 724 species/ 308 births & hatchings

Children's Zoo: 455 individuals of 96 species

GENERAL INFO

Hours: The Bronx Zoo is open 365 days a year. Hours: Mon-Fri, 10 am-5 pm; weekends and holidays until 5:30 pm. Nov-Mar, 10 am-4:30 pm every day.

Admission: General admission is charged (free on Wednesday). Children under 17 must be accompanied by an adult. Group rates are available for ten or more by calling (800) YES-2868.

Parking: Parking is available on site.

Visitor Rules: For the benefit of our wildlife and our guests: pets, radios, bicycles, skates, and skateboards are prohibited. Smoking is also prohibited, except in designated areas at Asia Gate (D9 on map) and Southern Boulevard Gate (A6 on map).

Website: www.wcs.org

← VICTORIA CROWNED PIGEON

INDEX OF ANIMALS

alligator, Chinese, 30-31, **85**, 96
antelope, 52
   blackbuck, 62
   blesbok, 50, 52
   nyala, 50, 52, **52**
   Thomson's gazelle, 50, 52
armadillo, 88, **88**
assassin bug, red-eyed, 38
babirusa (wild pig), 66, **66**
baboon, gelada, 42, **42**, **43**, 45, **45**, 47
   hamadryas, 104, **104**
bat, 67, 69, 103
   flying fox, 67
   Rodrigues fruit, 59, 69
   short-tailed fruit, **67**
   spear-nosed, 67
bear, grizzly, 70, **70**
   Kodiak, 70
   polar, 70, **70**, 103, **103**
   spectacled, 104
beetle, goliath, 38
binturong (bear cat), 55, **55**
bird of paradise, 76-78, **77**
bison, American, 52, 78, **79**, 8-99, **98**
boa constrictor, 30, 69, 88, **88**
caiman, broad-snouted, **31**
caribou, 98
cassowary, 22
catfish, upside-down, 38
cheetah, 48, **48**, 50
cock-of-the-rock, **76**
cockatoo, palm, 78, **78**, 96
condor, 26
cormorant, guanay, 21, **21**
crane, 78
   white-naped, 71
crocodile, dwarf, 38
currasow, helmeted, **77**, 78
deer, axis, 62
   barasingha, 62
   Formosan sika, 62, **63**
   Pere David, 74, **74**

Roosevelt elk, 104
sambar, 67
dolphin, Atlantic bottlenose, 103
domestic animals, 30
donkey, **27**
duck, 32-33
   mallard, 32, **32**, 33
eagle, North American bald, 26, **26**
elephant, African, 15
   Asian, **14**, 15, 58, 66, **66**
emu, 22
ferret, black-footed, 96
flamingo, 19, 24, **24**
fox, Arctic, 103
   fennec, 28, **28**
frog, 30, 38
   bullfrog, 28
   long-toed running, **33**
   poison dart, **30**
galago (bush baby), 69, **69**
gaur (wild cattle), 62, **62**, 66
gharial, **53**, 59, **59**
gibbon, white-cheeked, 22, 24, 56, **57**, 58
giraffe, 33, 48
   Baringo, 48, **49**
gorilla, Grauer's (eastern lowland), 42
   mountain, 38, 42, 93-94
   western lowland, **1**, 33, 38, **39**, **40**, **41**, 42, **83**, **95**, 96, 97
guanaco, 27, **27**
hawk, 24
heron, night, 28
hog, red river, 36, **38**
hornbill, 77-78
   great, **77**
horses, Mongolian wild (Przewalski's), 26, **26**, 94
hyrax, rock, **47**
ibex, Nubian, **44**, 45, 47
ibis, scarlet, 19, **19**
jaguar, 101
kangaroo, tree, **52**, 55, 58
leopard, 58
   black, 58, **58**
   snow, 71, **73**, **85**, 96-97, **96**

leopard cat, **68**, 69, **84**
lion, 50, **51**
lizard, 28, 38, 59
macaw, 94
manatee, West Indian, 91
mandrill, 33, **35**, 36, **90**
mantid, **58**
marmoset, 17
   pygmy, 17
   silvery, **17**
meerkat, 104
monkey, 17, 58, 93
   capuchin, **18**, 19
   colobus, 33, **34**
   DeBrazza's, 36, **37**
   Japanese snow, **103**
   proboscis, **54**, 55-56
   silvered langur, 56, **56**, 58
   Wolf's (guenon), 33, **36**
moose, 98
mormyrid, elephant-nosed, 38
mountain lion, 104, **105**
mountain sheep, 98
okapi, 33, **33**, 36
ostrich, 22
   Masai, **23**
otter, 28
   Asian small-clawed, 56, **56**
owl, 24, 88
   great horned, 28
panda, giant, 71, 93
   red, 71, **71**, 82, 103
parrot, 78
peafowl, 66
pelican, 22, 24
   white, **24**
penguin, chinstrap, 103
   gentoo, 103
   Magellanic, **20**, 21, 22
pheasant, 71-72, 78, 93, 101
pigeon, Victoria crowned, **108**
piranha, red-bellied, 103
puffin, tufted, 19
python, reticulated, 30
rhea, common, 22, **22**
rhinoceros, 58
   African black, 15, **15**, 58, 91
   Indian, **61**, 66-67
rodent
   chinchilla, 31, 32
   kangaroo rat, 28
   mole-rat, naked, 69, **69**
   mouse, Egyptian spiny, 32
      house, 32
      zebra, **32**
   prairie dog, **27**, 28, 31, 88, 104
   rat, Norway, 32
   squirrel, Prevost's, **32**
      southern flying, 32
scorpion, glow-in-the-dark, 69
sea lion, California, 16, **16**, 17, 103, **103**
seal, harbor, 103
sheep, Angora, **104**
snake, 28, 30, 38
spider, 28
stork, maribou, **50**
swan, black-necked, 32, **32**
   trumpeter, 72, **72**
tahr (wild goat), 67
tamarin, 17
   cotton-top, 104
   golden lion, **17**
tapir, Malayan, 15-16, **15**, 58
tarantula, 28
tern, Inca, 21, **21**
tiger, 93
   Siberian, **64-65**, 66, **66**, **94**
toad, marine, 28
tortoise, 28, **88**
   radiated, 31, **31**
tragopan, Temminck's, 71-72, **72**
turtle, 30, 38
vulture, 26
   king, **25**
walking stick, 28, 59, **59**
wallaby, 28, **29**, 104
walrus, 103, **103**
whale, beluga, 102, 103
wolf, Mexican gray, 74, **75**
zebra, 33, 50, 52, **52**

 # in bold = photo

PHOTO CREDITS

William Conway: 92

William Karesh: 90

Wildlife Conservation Society:

Dennis DeMello: 1, 4, 16 (close-up), 20, 21, 22, 23, 24 (flamingos), 25, 26 (eagle), 27 (guanaco), 28, 29, 31 (caiman), 32 (mallard), 33 (okapi), 36, 37, 38, 39, 40, 41, 43, 46, 48 (cheetahs), 49, 51, 52, 53, 54, 56 (langur), 57, 58, 59, 60, 61, 63, 66 (tiger), 69 (naked mole-rat), 70 (grizzly), 72, 73, 75, 76, 77 (chicks, hornbill, currasow), 78, 82, 83, 84, 85 (leopard), 86, 87, 88, 89, 94, 95, 102, 105, 107, 108

Elyssa Dickstein: 17 (tamarin), 18, 34, 42, 45, 79, 103 (walrus), 104 (sheep)

Historical images: 97, 98, 100

Bill Meng: cover, 2-3, 14, 15, 16 (group), 17 (marmoset), 19, 24 (pelican), 30, 32 (squirrel, mouse, swan), 33 (frog), 35, 44, 48 (giraffe), 55, 56 (otter), 62, 64-65, 66 (babirusa, elephant), 67, 68, 69 (galago), 71, 72, 74, 77 (nest, adult), 85 (alligator), 96, 103 (sea lion, bear), 104 (baboon)

Diane Shapiro: 8, 26 (horses), 27 (donkey, exhibit), 31 (tortoise), 47 (visitors), 50, 70 (polar bear), 91, 106

Susan Suarez: 47 (hyrax)

Tom Veltre: 93